# Evolutionary Biology

VOLUME 32

A Continuation Order Plan is available for this series. A continuation order will bring delivery of each new volume immediately upon publication. Volumes are billed only upon actual shipment. For further information please contact the publisher.

## Limits to Knowledge in Evolutionary Genetics

# Evolutionary Biology

## VOLUME 32


**Edited by**

**MICHAEL T. CLEGG**
*University of California, Riverside*
*Riverside, California*

**MAX K. HECHT**
*Queens College of the*
*City University of New York*
*Flushing, New York*

**and**

**ROSS J. MACINTYRE**
*Cornell University*
*Ithaca, New York*



KLUWER ACADEMIC / PLENUM PUBLISHERS
NEW YORK, BOSTON, DORDRECHT, LONDON, MOSCOW

The Library of Congress catalogued the first volume of this title as follows:

Evolutionary biology, v. 1–     1967–

New York, Appleton-Century-Crofts.
v. illus., 24 cm annual.
Editors: 1967–     T. Dobzhansky and others.
1. Evolution—Period. 2. Biology—Period. I. Dobzhanksy, Theodosius Grigorievich, 1900–

QH366.A1E9     575′.005     67-11961

ISSN 0071-3260

ISBN: 0-306-46227-3


233 Spring Street, New York, N.Y. 10013

http://www.wkap.nl/

10 9 8 7 6 5 4 3 2 1

A C.I.P. record for this book is available from the Library of Congress



Printed in the United States of America

# Contributors

**Charles F. Aquadro** • *Department of Molecular Biology and Genetics, Cornell University, Ithaca, New York 14853*

**Stephane Boissinot** • *Laboratory of Molecular and Cellular Biology, National Institute of Diabetes and Digestive and Kidney Diseases, National Institutes of Health, Bethesda, Maryland 20892-0830*

**Gary A. Churchill** • *The Jackson Laboratory, Bar Harbor, Maine 04609*

**Andrew G. Clark** • *Institute of Molecular Evolutionary Genetics, Department of Biology, Pennsylvania State University, University Park, Pennsylvania 16802*

**Michael T. Clegg** • *Department of Botany and Plant Sciences, University of California, Riverside, California 92521*

**David Hewett-Emmett** • *Human Genetics Center, University of Texas, Houston, Texas 77225*

**Jody Hey** • *Department of Genetics, Nelson Biological Labs, Rutgers University, Piscataway, New Jersey 08854-8082*

**Wen-Hsiung Li** • *Department of Ecology and Evolution, University of Chicago, Chicago, Illinois 60637*

**Michael Lynch** • *Department of Biology, University of Oregon, Eugene, Oregon 97403*

**Leonard Nunney** • *Department of Biology, University of California, Riverside, California 92521*

**Alex Rosenberg** • *Department of Philosophy, Duke University, Durham, North Carolina 27706*

**Michael Ruse** • *University of Guelph, Guelph, Ontario, N1G 2W1, Canada*

**Song-Kun Shyue** • *Institute of Biomedical Sciences, Academia Sinica, Taipei 11529, Taiwan*

**Ying Tan** • *Human Genetics Center, University of Texas, Houston, Texas 77225*

**Ward B. Watt** • *Department of Biological Sciences, Stanford University, Stanford, California 94305-5020; and Rocky Mountain Biological Laboratory, P.O. Box 519, Crested Butte, Colorado 81224*

**Bruce S. Weir** • *Program in Statistical Genetics, Department of Statistics, North Carolina State University, Raleigh, North Carolina 27695-8203*

**Chung-I Wu** • *Department of Ecology and Evolution, University of Chicago, Chicago, Illinois 60637*

# Preface

This collection of chapters is the result of a project funded by the Alfred P. Sloan Foundation on the "limits to knowledge." The Sloan Foundation has funded several projects to explore limits to knowledge in various areas of science, and I was privileged to acquire funding from the foundation to explore limits to knowledge in evolutionary biology. My particular approach to this interesting question was to organize a small meeting with a number of prominent evolutionary geneticists and philosophers of science, held in January 1998, in Riverside, California. The participants were asked to bring an essay on the question of limits to knowledge to the meeting, and this volume is the revised and edited fruit of that endeavor.

Why consider the notion of limits to knowledge? Some of the participants in the Riverside meeting, and many others, have questioned the value of even asking whether there are limits to knowledge in evolutionary biology. Some are concerned that considering the question of limits somehow might give ammunition to the creationist movement and thereby weaken public and scientific support for the science of evolution. This is certainly a justifiable concern when political and religious doctrines are sometimes cloaked as science and presented to educators for inclusion in science curricula under a disingenuous equal-time argument. I believe the answer to this objection is eloquently addressed in the essay by the philosopher and historian of biology, Michael Ruse (Chapter 1), who makes a crucial distinction between the scientific fact of evolution and the scientific problem of reconstructing its paths and causes. There is no dispute about the fact of evolution among the authors of these chapters, or indeed among the vast majority of scientists engaged in understanding the history of the earth and of the life forms that occupy it.

Even granting that a study of limits does no damage, is there value in addressing this topic? I believe there are two positive answers to this question. The first is that a search for limits should help illuminate the fruitful questions taken from those biologically interesting questions that are presently (or perhaps permanently) outside the reach of science. The nature of science is to work on the boundaries between the known and the unknown. These boundaries shift as new methods are developed (e.g., the invention of the microscope, or more recently the elaboration of rapid DNA sequencing technologies) and as new concepts are elaborated (e.g., the theory of the gene, or more recently, the coalescence framework in population genetics). These new tools allow us to address a set of interesting questions that previously were outside the realm of science; as a consequence, the boundary between the knowable and the unknowable has shifted. I contend that a study of limits should reveal and clarify the boundaries and thereby make sharper the set of fruitful questions.

My second reason concerns the place of science in major public policy debates. We are presently witnessing a major debate about global warming, where the methods of science and the limits of science are sometimes misunderstood by public policymakers. I was stimulated to consider the limits to knowledge in the applied science of conservation biology as Chair of the National Research Council Committee on Science and the Endangered Species Act. The theoretical foundations of ecology and population genetics play a major role in the science of biological conservation. For instance, theory, in conjunction with empirical information, informs approaches to reserve design and to the elaboration of breeding strategies. The fundamental science of evolutionary biology is called on by managers in deciding when a species is no longer in danger of extinction and when a listing is the appropriate action. The list of places where scientific information is applied in this field could be extended indefinitely. So we can conclude that knowledge in evolutionary biology is used in manifold ways to solve practical problems. What is intriguing and often underappreciated is that our knowledge base is often approximate, conditional, and subject to revision as the dialogue of science advances. Policymakers and the society they serve expect science to provide answers to problems that may transcend the boundaries of scientific knowledge. A frank exploration of these boundaries seems highly appropriate.

Let me now address the title of this volume. In the course of this project, I came to realize that evolutionary biology is too large and too diverse a topic for the set of chapters that are presented here. I am a population geneticist who also works in molecular evolution. The authors of the chapters presented in this volume reflect my particular interests, and it is clear that the center of gravity is heavily weighted toward population and quantitative genetics. Accordingly, the volume is given the more restricted title "Limits to Knowledge in Evolutionary Genetics."

I want to thank the Alfred P. Sloan Foundation and Dr. Jesse Ausubel for their support of this project. I thank my colleagues, Dr. Michael Cummings and Dr. Mary Durbin, for their ideas and contributions to this effort. I also thank Ms. Cindi McKernan for assistance with the Riverside meeting,

Ms. Laura Heraty for help with the final preparation of the manuscripts, and the Department of Botany and Plant Sciences at the University of California, Riverside, for providing unstinting assistance. Finally, I am deeply indebted to the participants who enthusiastically explored and critiqued the idea of limits and who patiently wrote and revised their chapters for this volume.

Michael T. Clegg
Riverside, California

# Contents

## II. The Philosophy of Biology, Paradigms, and Paradigm Shifts

## III. Limits to Historical Inference and Prediction

### 6. Inferring Ancestral Character States

*Gary A. Churchill*



### 7. The Problem of Inferring Selection and Evolutionary History from Molecular Data

*Charles F. Aquadro*

## IV. Quantitative Genetics and the Prediction of Phenotype from Genotype

# Evolutionary Biology

VOLUME 32

# I

# Limits to Knowledge

## An Introduction

Are there limits to knowledge in evolutionary biology? The last 25 years have witnessed an explosion in biological knowledge. Much of this explosion has been driven by an unparalleled pace of technological innovation. Today we have the tools to answer questions that we could only speculate about a generation ago. It is not so much that new theoretical constructs have arisen as it is that our tools for empirical exploration have expanded enormously. If this history is any guide to the future, we can expect to see a continuing elaboration of empirical knowledge. Indeed, the current genomics enterprise represents one of the great descriptive phases of biology, perhaps akin to the age of discovery two centuries ago. So it is easy to question the notion of limits altogether. But that would be too facile. A deeper analysis suggests that there are classes of problems about which our knowledge is at best approximate and incomplete. To set the stage for an analysis of limits, the historian and philosopher of biology, Michael Ruse (Chapter 1) reminds us that evolutionary science has traditionally addressed three problems. The first, and largely completed problem, has been to establish the fact of evolution. This was done, beginning with Darwin, by using the consilience of different lines of evidence from disparate areas of investigation to establish an overwhelming case for the fact of evolution. The second problem area is historical inference with which we seek to understand the evolutionary pathways that led to today's biological world. It is clear that precise knowledge of historical pathways is limited and will remain limited. (Several of the chapters in later sections of this volume explore this question of historical inference from different perspectives.) The third area concerns the mechanisms of evolutionary change. This is an area where theoretical research is still exploring mechanisms involving, for example, models of social behavior, sexual selection, the evolution of recombination, and macroevolution.

To these three, I would add a fourth area, which is prediction. Evolutionary science is applied in plant and animal breeding, human health, and conservation biology for prediction. There are well-formulated statistical approaches to predict, for example, the response to selection in breeding populations or the minimum viable population size in conservation biology. It is clear that our ability to predict is limited by the same stochastic forces that limit historical inference. Ruse goes on to explore the influence of social forces on the kinds of questions we consider important in contemporary biology. Here, social context is seen as a kind of limitation on knowledge because what the community sees as being worthy of study may be a consequence of the social context of the times in which we live. The high-priority questions of a different era or a different society would likely diverge substantially from those that intrigue our contemporary western society. In Chapter 2, I take a more narrow approach and consider how the theoretical development of population genetics may have led to empirical questions that are misleading, either because the fundamental parameters of population genetics are not estimable, or because the simplifications of theory remove some of the essential details of evolutionary change. I also consider the problem of adaptation as a central empirical question that is difficult to penetrate because the process plays out over several different levels of biological organization. These two chapters set the stage for the ideas explored throughout this volume.

1

# Limits to Our Knowledge of Evolution

MICHAEL RUSE

## INTRODUCTION

In this discussion I want to raise the question of whether there are limits to our knowledge of evolution, and if so of what form? Are we barred from full understanding by irretrievable loss of information, or by a complexity of a degree that cannot be untangled, or by more profound and far-reaching factors pertaining to the very nature of the scientific enterprise itself? To structure my discussion I will make the customary threefold distinction between the *fact* of evolution, the *path* or paths of evolution, and the *theory* or mechanisms or causes of evolution (Ayala, 1985; Ruse, 1982, 1984). It is not always possible to adhere rigorously to this trichotomy, nor is it necessary, but it is a useful way of keeping things clear and separate.

## THE FACT OF EVOLUTION

By the "fact" of evolution I mean the central claim put forward by Charles Darwin in his *The Origin of Species* (1859): That all organisms, living and dead, originated from other forms by natural, that is to say law-bound, processes of generation and development. Usually one includes here the idea that the original forms were somewhat if not significantly simpler

MICHAEL RUSE • University of Guelph, Guelph, Ontario, N1G 2W1, Canada.

*Evolutionary Biology, Volume 32*, edited by Michael T. Clegg *et al.*
Kluwer Academic / Plenum Publishers, New York, 2000.

than the final forms, those living today (Ruse, 1996). One question of some debate is whether the fact of evolution should stretch back to include the origination of living forms from nonliving elements. As it happens, at least in print, Charles Darwin himself was silent on this matter. But today, for most people the fact of evolution would include some such event or events (Ruse, 1982; Dawkins, 1986; Strickberger, 1990; Ridley, 1993; Freeman and Herron, 1998).

What is the status of the claim that evolution as fact occurred? The answer surely is that this indeed is something within the limits of our knowledge, and (leaving aside for one moment the inorganic–organic transition) is about as secure a proposition as it is possible to have in science. The case made by Darwin in *The Origin* is still definitive. Most particularly, he offered a consilient argument, an idea recently made popular by the evolutionist Edward O. Wilson (1998) but going back to the writings of Darwin's contemporary, the historian and philosopher of science, William Whewell (1840). Darwin argued that the truth of a claim can be established beyond reasonable doubt if it is possible to show that the claim is supported by evidence from many different areas of study, and conversely throws explanatory light on these various areas (Ruse, 1979a) (Fig. 1). Darwin (1859) argued that the fact of evolution is supported by phenomena right through the organic world, and in turn explains them. Why, for instance, is there the peculiar distribution of birds and reptiles in the Galápagos Archipelago, with very similar but slightly different forms from island to island, and with such forms similar but likewise different from the South American mainland (Fig. 2)? Why is there the gradual progressive rise to be seen in the fossil record, from strange and simple forms to the remains of complex organisms only slightly different from those still extant (Figs. 3, 4)? Why does one find similarities (homologies) between the skeletons of the forelimbs of mammals, yet the limbs themselves are used for quite different functions (Fig. 5)? Why are there frequently similarities also between the embryos of animals, which as adults are very different (Fig. 6)? Why do organisms fall into the structured branching pattern shown by the Linnaean hierarchy, rather than show a different pattern, or no pattern at all? All these "whys." Evolution explains each and every one of these phenomena. Conversely, these phenomena in turn all point to the truth of the fact of evolution. The birds and reptiles of the Galápagos came to the islands from the mainland and evolved as they moved from island to island. The fossil record shows the progressive upward evolutionary rise of organic life. The mammalian forelimbs reveal their shared ancestry, and so forth. The phenomena of the organic world are the clues—the bloodstains, the torn fabric, the broken alibi, the fingerprints—that point unambiguously to the cause or culprit, namely evolution (Ruse, 1973).

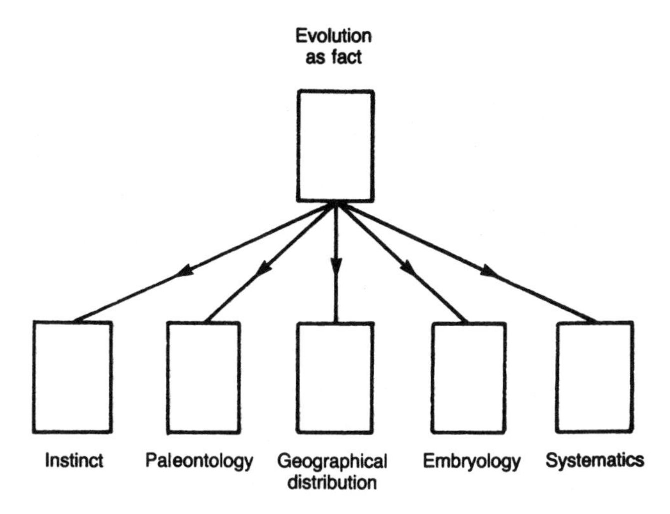


FIG. 1. The structure of Darwin's argument for the fact of evolution. The fact explains and unifies claims made in the subdisciplines (only some of which are shown), which later in turn yield the "circumstantial evidence" for the fact itself.

It was this unificatory argument that convinced Darwin's contemporaries and still convinces us today. Evolution as a fact is established—within the bounds of certain knowledge—thanks to the overwhelming circumstantial evidence that points to its truth. I say this notwithstanding the fact that today, particularly in the nonscientific community, this is a conclusion that is much disputed. There are many who would argue that evolution in all forms is false, or at the very least sufficiently hypothetical to be labeled a mere "theory," rather than a fact (Plantinga, 1991a,b; Johnson, 1991, 1995). The use of "theory" here meaning "very iffy," as in "I have a definitive theory about Kennedy's assassination," as opposed to "theory" meaning "causal hypothesis," as in "Newton gave the theory behind the Copernican Revolution" (Scott, 1997).

Here let me say simply that although such arguments by the critics are politically dangerous—an ongoing and serious threat to the proper education of our children—they are invariably based on ignorance, or religious prejudice, or a combination of both (Ruse, 1988a,b, 2000). Those who argue that evolution is only a theory and not a fact never acknowledge the consilient nature of the Darwinian argument in favor of evolution, and invariably they confuse the fact of evolution with the path of evolution. They concentrate on supposed gaps in the fossil record, showing that paths are not always readily traceable, and then they project this ignorance back onto

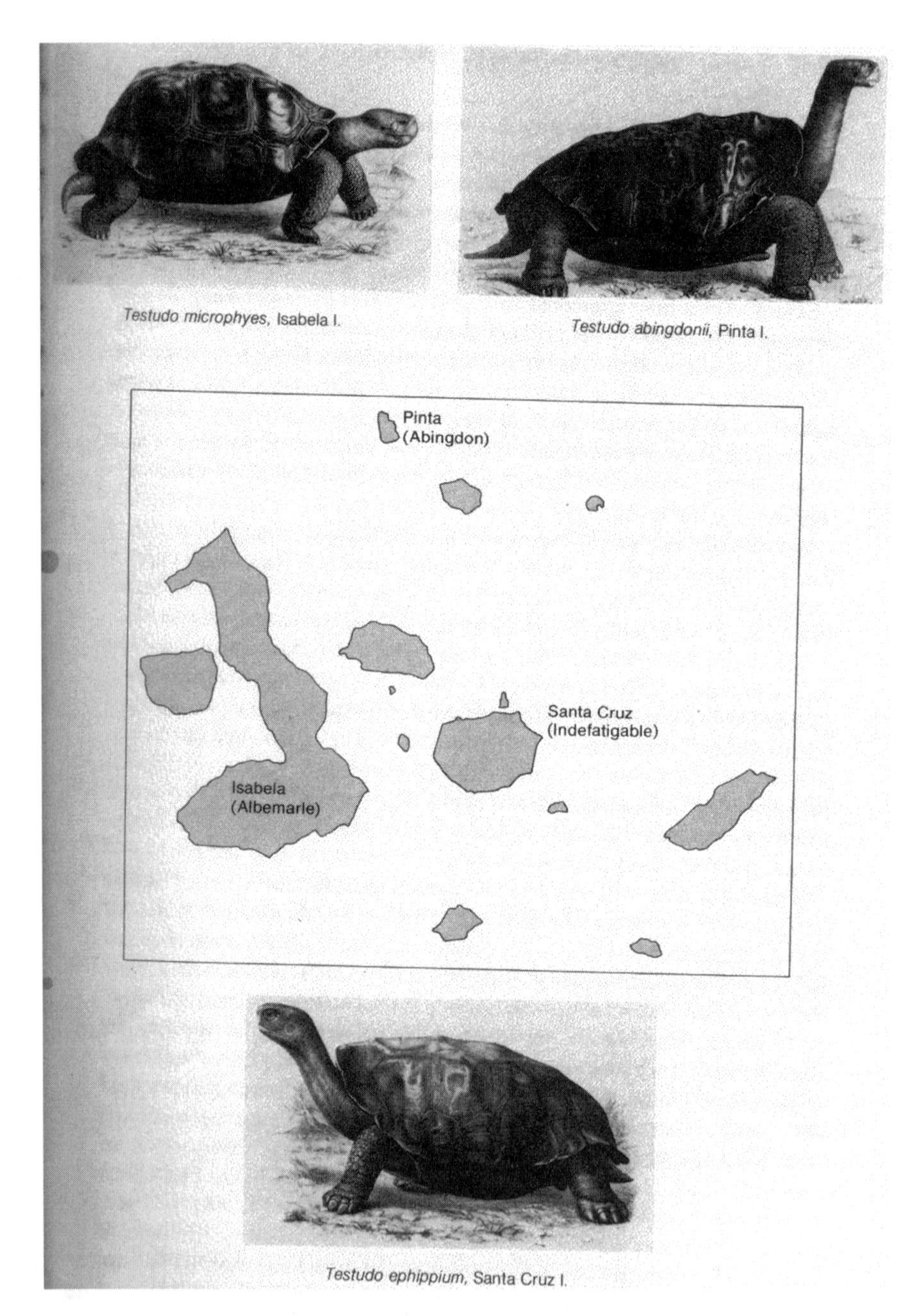


FIG. 2. Three different tortoises from three different islands of the Galápagos. (From Dobzhansky *et al.*, 1977; © 1997 by W. H. Freeman and Company. Used with permission.)

TABLE of STRATA and Order of Appearance of Animal Life upon the Earth.

| Era | Strata | Formation | Order of Appearance of Animal Life | |
|---|---|---|---|---|
| TERTIARY or CÆNOZOIC | Turbary. Shell-Marl. Glacial Drift. Brick Earth. Bone-Caves. | Pleistocene | MAN by Remains. by Weapons. | Birds and Mammals. |
| | Norwich, Red, Coralline } Crag. | Pliocene | | |
| | Faluns. Molasse. | Miocene | Ruminantia. Quadrumana. Proboscidia. Birds, Orders of. | |
| | Gyps. London, Plastic } Clays. | Eocene | Rodentia. Ungulata. Carnivora. Mammals Orders of. | |
| SECONDARY or MEZOZOIC | Maestricht. Upper Chalk. Lower Chalk. Upper Greensand. Lower Greensand. | Cretaceous | Cycloid., Ctenoid. } Fishes. Mosasaurus. Polyptychodon. Birds, by Bones. Procœlian Crocodilia. | |
| | Weald Clay. Hastings Sand. Purbeck Beds. | Wealden | Iguanodon. Marsupials, — Chelonia by Bones. | |
| | Kimmeridgian. Oxfordian. Kellovian. Forest Marble. Bath-Stone. Stonesfield Slate. Great Oolite. Lias. Bone Bed. | Oolite (U., M., L.) | Pliosaurus. Marsupials. Icthyopterygia. Amphicœlian Crocodilia. Pterosauria. Homocercal Fishes. Cephalopods 2-gilled. MAMMALIA. | Reptiles. |
| | U. New Red Sandstone. Muschelkalk. Bunter. | Trias | AVES, by Foot-prints. Sauropterygia. Labyrinthodontia. Crustacea 10-poda. | |
| PRIMARY or PALEOZOIC | Marl-Sand. Magnesian Limestone. L. New Red Sandstone. | Permian | Sauria. Chelonia, by Foot-prints. Isopoda. | |
| | Coal-Measures. Mountain Limestone. Carboniferous Slate. | Carboniferous | REPTILIA ganoceph. Insecta | Fishes. |
| | U. Old Red Sandstone. Caithness Flags. L. Old Red Sandstone. | Devonian | PISCES { ganoid. placo-ganoid. placoid. Heterocercal. | |
| | Ludlow. Wenlock. Caradoc. Llandeilo. Lingula Flags. | Silurian | Echinoderms. Annelids. Bivalves. Trilobites. Pteropods. Brachiopods. Gastropods. Cephalopods 4-gilled. | Invertebrates. |
| | Cambrian. | | Fucoids. Zoophytes. | |

FIG. 3. The fossil record as known at the time of *The Origin of Species*. By this time, all serious scientists knew that the record involved much branching. This diagram is taken from Richard Owen's *Paleontology* (1861), but everyone, evolutionist or not, accepted the same basic pattern. Of course, nothing was known of actual dates, but if one compares this with a modern-day account, the picture is surprisingly similar.

| Era | Period | | Epoch | Events |
|---|---|---|---|---|
| Cenozoic | Quaternary | | Pleistocene | Evolution of Man |
| | Tertiary | | Pliocene<br>Miocene<br>Oligocene<br>Eocene<br>Paleocene | Mammalian Radiation |
| Mesozoic | Cretaceous | | | Last Dinosaurs<br>First Primates<br>First Flowering Plants |
| | Jurassic | | | Dinosaurs<br>First Birds |
| | Triassic | | | First Mammals<br>Therapside Dominant |
| Paleozoic | Permian | | | Major Marine Extinction<br>Pelycosaurs Dominant |
| | Carbo-niferous | Pennsylvanian | | First Reptiles |
| | | Mississippian | | Scale Trees, Seed Ferns |
| | Devonian | | | First Amphibians<br>Jawed Fishes Diversify |
| | Silurian | | | First Vascular Land Plants |
| | Ordovician | | | Burst of Diversification in Metazoan Families |
| | Cambrian | | | First Fish<br>First Chordates |
| Precambrian | Ediacaran | | | First Skeletal Elements<br>First Soft-Bodied Metazoans<br>First Animal Traces (Coelomates) |

Millions of Years Ago: 0, 50, 100, 150, 200, 250, 300, 350, 400, 450, 500, 550, 600, 650

FIG. 4. Major events in the evolution of multicellular life. (From Valentine, 1978. Reprinted by permission of the artist, Patricia J. Wynne.)

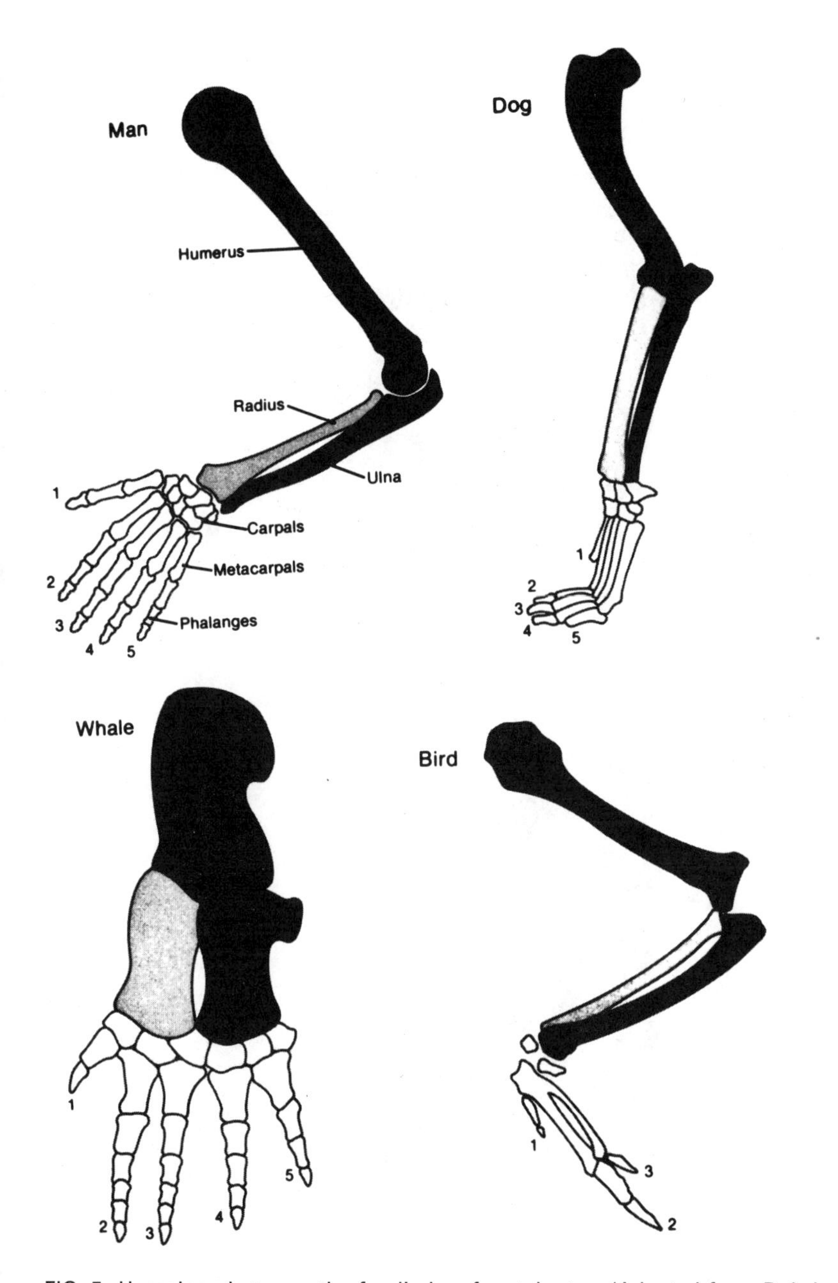


FIG. 5. Homology between the forelimbs of vertebrates. (Adapted from Dobzhansky *et al.*, 1977; © 1997 by W. H. Freeman and Company. Used with permission.)

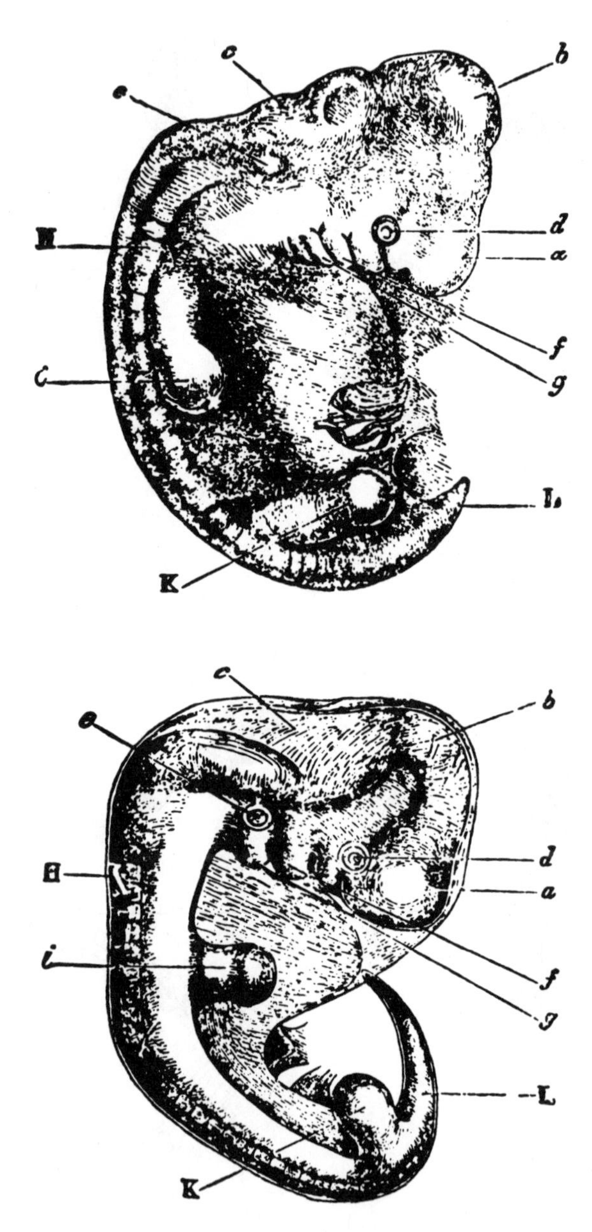


FIG. 6. Comparison of human and canine embryos. (From Charles Darwin's *The Descent of Man*, 1871.)

the very fact of evolution itself. The error of such an approach is shown clearly by the fact that even if we were ignorant entirely about the paths of evolution—if the fossil record were a totally closed book—evolution as fact would still stand, thanks to such things as the peculiar distributions of the birds and reptiles in the Galápagos and the already-mentioned homologies between the four limbs of the vertebrates. A confession is not needed to establish guilt "beyond reasonable doubt" nor is the fossil record (assuming for the sake of argument that it is this which is the evidential equivalent of a confession) needed to establish evolution "beyond reasonable doubt."

One antievolutionary argument that has received much attention recently, given authority since it comes from a practicing research scientist, is that of the biochemist, Michael Behe (1996). He argues that in many respects life has such a degree of complexity that it cannot possibly be explained by natural law. In the light of such "irreducibility," evolutionary explanations are precluded. In support of his case, Behe offers the detailed example of the blood-clotting mechanism in mammals: he argues that it is of such a kind that, were any part of the mechanism removed, the whole would cease to function. Hence, it is impossible that the parts have evolved separately, being then assembled by natural forces. Behe draws an analogy with a mousetrap that has some five functioning parts, any one of which is necessary and the removal of which would destroy the functioning of the trap. Behe argues that the mousetrap shows some kind of intervening non-natural design and the same must be true of such complex phenomena as mammalian blood clotting.

Let me say simply that this claim seems not to be well taken, either as science or as theology for that matter. On the one hand, there simply is no evidence that organic characteristics are so complex that they could not have been produced by an evolutionary process. In the case specifically of blood clotting, the world expert on the subject, Russell Doolittle (1997), has pointed out (in response to Behe) that far from the parts of mammalian blood-clotting process being so intricate that were one piece removed everything would collapse down, there are now artificially engineered organisms that function perfectly well with two or more parts of the clotting removed. What Behe fails to understand is that the present-day state of an organism is not necessarily a direct function of its evolutionary past. It may well have been that in the past certain elements were there, bridging gaps: subsequently, the gaps have been filled in by other parts still existing, whereas the original elements are now missing. While it may indeed be true that were one to try to disassemble parts as they exist at the moment, everything would collapse down into nonfunctionality, it is not necessarily the case that this is the way in which such complex organisms were built in the first place. And in any case, there is something highly suspicious about this appeal to the "God of the gaps." At the very least, what one is doing is taking the discussion out of the scientific domain and suggesting that there are areas in which science cannot encroach: a somewhat risky enterprise, given past history. Many people (usually philosophers) "proved" that you will never explain life in terms of the molecules, and then along came Watson and Crick.

The much-cited critic of evolution, the Berkeley lawyer Phillip Johnson (1991, 1995), takes a somewhat different tack. He argues that evolution is less a fact in the sense of something empirically established and has status more as a kind of metaphysical precommitment to the study and understanding of organisms. Johnson argues that evolutionists start with a prescientific commitment to naturalism, that is to say, to explanation through unbroken law without reference to nonnatural (for instance divine) causes. But given such a commitment, argues Johnson, evolution follows as an inevitable consequence. If one is a naturalist of some form, then necessarily one is an evolutionist, since presumably the only way in which one can explain organisms is as a consequence of an evolutionary process.

However, this argument is surely not well taken. Even if it were true (which I quite happily concede) that the evolutionist is in some form a naturalist (methodologically, at the very least), it is surely not the case that the naturalist is necessarily committed to evolution. One could, for instance, argue that organisms in some sense spontaneously form themselves full-blown from inorganic material, much in the way that the Greek anatomists supposed organisms to appear. If, for instance, lions and tigers just formed by natural law from mud, this would be very different from what we today have in mind by "evolution as fact." But more than this, one might suppose that nature is eternal in some sense, and that life forms themselves are eternal and unchanging. That is to say, not only are planets and suns things that go back indefinitely in time, but so also are lions and tigers. As it happens, we do not think that any of this is true; but from the mere assumption of naturalism, one cannot disprove these hypotheses. And certainly, if one held to such beliefs, one would hardly be an evolutionist. Hence, one can see that naturalism and evolutionism do not have a direct connection and the one does not follow at once from the other.

Having said this—recognizing that half-truths are often much more effective than outright fabrications—let me concede that there is something in what Johnson says. As things stand today, one has to admit that there is a gap in our understanding of the natural emergence of life from nonlife. I am not sure whether it would be correct to say that such a belief (in the natural emergence of life from nonlife) is as much a reflection of naturalism as of hard scientific evidence, but if someone were to insist that this is the case, it would be difficult to refute them at the moment (Ruse, 1997). I would not exaggerate. We now have a good idea of when life supposedly did emerge here on earth (rather more than 3.5 billion years ago, as soon as the earth cooled enough to bear life), we have knowledge of how inorganic materials can be put together to make organic materials (Miller, 1953, 1992), and we have various hypotheses about how life might have formed (Fox, 1988). But I would have said that we cannot as yet absolutely rule out alternative hypotheses such as that the earth was "seeded" with primitive life from elsewhere (Crick and Orgel, 1973), whereas we can surely rule out the hypothesis that the earth was populated with elephants and tigers from elsewhere. Presumably under such an alternative hypothesis as this, given the big bang theory of the universe's origin, one must assume that life started somewhere sometime, and as a naturalist one assumes that this was natural. But one cannot as yet state as established fact that life started here on earth.

This is not to say that the origin of life is beyond the limits of our knowledge. One supposes that if one could produce life experimentally and show that the initial conditions were precisely such as obtained back in time on this earth, one would have gone far to make the case for the naturalistic origin of life on earth. But this has not yet been done. Not that this failure gives the critic like Johnson the conclusion he wants, that belief in life's natural origin is a "religious" belief with the same epistemological status as the Genesis story of Creation. Belief that the world works according to law is one that is well justified by the success of all science, and not just evolutionary science. It is not a leap of faith in the sense of going in the face of evidence or unreasonably beyond the evidence. It is in a sense precisely what we mean by being reasonable, for assuming naturalism has led to one explanatory triumph after another. Indeed, even as I write and you read, there are researchers at the interface of chemistry and biology racing to put together, through naturally simulating processes, macro-organic molecules of a kind that truly could be considered in the realm of the living. James Ferris and co-workers (Ferris *et al.*, 1996) have grown long chains of amino acids using clays as substrates [as long suggested by A. G. Cairns-Smith (1982, 1986)], and David Bartel and co-workers (Bartel and Szostak, 1993; Ekland *et al.*, 1995; Ekland and Bartel, 1996) are trying to make self-replicating RNA. So even the most religious of people would be wise to hedge their bets. But this said, even though there is the risk that critics will misinterpret and twist what is being said, one should not today pretend that a naturalistic understanding of the ultimate origin of life is more detailed and confirmed than it truly is.

## THE PATH OF EVOLUTION

Move now to the second part of our inquiry, that of the status of claims about the path or paths of evolution. Claims in this domain have always been problematic and continue to be so to this day. It is an interesting if paradoxical fact that the first biologists to work on the history of organic life were, if anything, those most opposed to evolutionary speculations. One thinks particularly of the early 19th-century French biologist, the father of comparative anatomy, Georges Cuvier (Bowler, 1976; Rudwick, 1972; Coleman, 1964). It was he who first established securely the roughly progressive nature of the fossil record, from primitive invertebrates up to mammals of a kind that we see around us today. Yet he did this while denying the possibility of any genuine transitions from one form to another. To Cuvier, organisms as integrated functioning entities simply could not have come about through natural processes of any kind, and most particularly not through natural evolutionary processes of any kind (Ruse, 1979a).

Following Cuvier, knowledge of the fossil record continued to improve; with the acceptance of evolution as fact, Cuvier's doubts were dismissed and the record was interpreted as evidence of evolution, thanks particularly to Darwin's great metaphor, as the remains of a great branching tree of life showing an evolution from the simple to the complex, from "monad to man." In the post-*Origin* period, huge amounts of effort were devoted by evolutionists to uncovering the fossil record, most particularly by the great American paleontologists who raced against one another in the opening West to uncover the fabulous vertebrates of past times (Rainger, 1991; Ruse, 1996). At the same time, 19th-century evolutionists turned to other areas of the biological world, most particularly embryology, for help in discerning organic histories. Notorious was the so-called "biogenetic law" of the German evolutionist Ernst Haeckel, who argued that ontogeny, that is to say the development of the individual organism, recapitulates phylogeny, that is to say the development of the organism's past race or line (Richards, 1992).

Yet for all the successes—and there were many, most famously deciphering the past history of the horse—by around the turn of the century realization had come flooding in that the finding of phylogenies would be nothing as simple or straightforward as people had once optimistically supposed. It became clear that, for all the wonderful discoveries, there were still massive gaps in our knowledge of the record; and complementing this, it was increasingly obvious that all too frequently phylogeny and ontogeny are things apart. It is a very dangerous assumption indeed to suppose that the developing embryo necessarily goes through the past history of its ancestral race. It seems fair to say indeed that so dubious was the activity of phylogeny tracing that many of the best young biologists, men like E. B. Wilson and T. H. Morgan, turned in disgust from evolutionary studies to more profitable areas, such as cytology and the newly developing science of genetics (Allen, 1978; Maienschein, 1991).

Bringing the story rapidly down to the present, today we are obviously in a better situation than were biologists at the beginning of the century. Yet, one does not have to be unduly cynical to question whether things have improved to the extent that one would now say our knowledge of evolution's past was of the standing of our acceptance of the fact of evolution, or whether the prospects are that we will ever get a full understanding of the paths or course of evolution. Some phylogenies will forever be beyond the limits of our understanding and knowledge. It is true that in some areas we now have very detailed knowledge (based on fossil evidence) of the way in which organisms evolved. It is surely the case that the past 5 million years of human evolution, if not complete by any means, is looking really quite strong: taking us from *Australopithecus afarensis* down to *Homo sapiens*, with various side branches. We now can answer all sorts of questions that were hidden from view before, for instance, the notorious question about whether or not human brains evolved first and then we got up on our hind legs [as Darwin (1871) thought] or whether the converse happened. *Australopithecus afarensis*, particularly as represented by "Lucy," shows clearly that humans became bipedal before their brains started to explode up to the size that they have today (Johanson and Edey, 1981). And there are other areas of the fossil record where we likewise have good and ever-improving understanding, for instance, the transition from land mammals back to sea mammals like the whale (Carroll, 1997) (Fig. 7).

Combined with more knowledge of the fossil record, we also have new methods of discerning the past: for instance, the use of molecular tools and theories. The so-called theory of molecular drift provides us with a useful molecular clock, whereby we can measure the times since the ancestors of living organisms broke off from main stems and the like (Ayala, 1978). In addition to these new tools, we have refurbished old tools, particularly those utilizing embryology and comparative anatomy and the like. Based on increased sensitivity to questions about when development and its effects can and cannot tell about history, the philosophy of taxonomy known as "cladism" has virtually transformed the discerning of relationships and their past history (Ridley, 1986). It is perhaps true that cladism is not always quite as new as its most enthusiastic proponents suggest; but even critics must allow that phylogeny tracers have more powerful tools at hand than before (Sober, 1988), if only because new technology has transformed much that

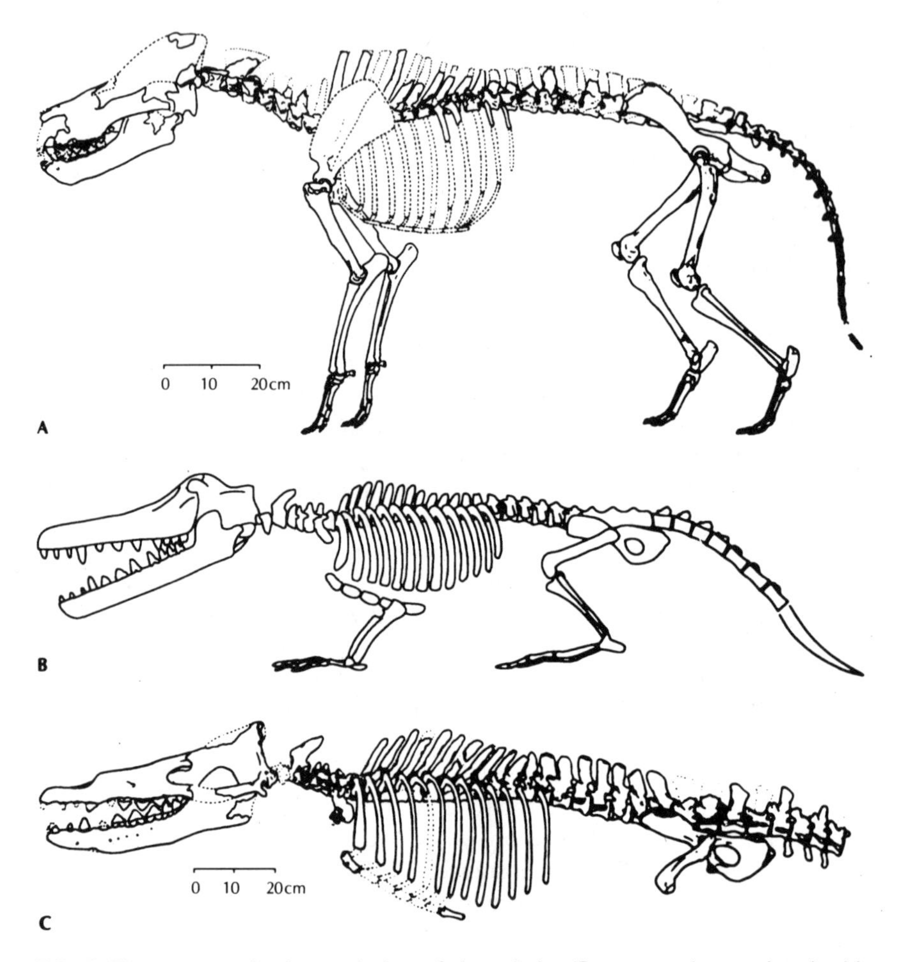


FIG. 7. Three stages in the evolution of the whale. (Parts a and c reprinted with permission from *Nature*. © 1994 Macmillan Magazines Limited. Part b reprinted with permission from *Science*.) *In R. L. Carroll, Patterns and Processes of Vertebrate Evolution, 1997.*

was done hitherto by hand and by brute brain power. One thinks particularly of the coming of computers, which enable students of the past to make much more powerful analyses involving far greater amounts of empirical information in far less time.

Yet when all is said and done, no one could claim that our knowledge of the past is complete or that probably it will ever be complete. Here indeed we do run into limits to our understanding. Even such basic questions as whether or not the birds evolved directly from the reptiles or via that particular branch of reptiles known as dinosaurs is not yet a finally settled question (pro: Ostrom, 1976; con: Feduccia, 1996). It is probably true to say that in particular instances the general consensus is toward one solution rather than another (most people now think that the birds evolved from the dinosaurs), but there are usually articulate and respected figures in the field who would argue otherwise. When we turn to branches of inquiry where the fossil record gets weaker and weaker, it is surely the case that not only is our knowledge incomplete now but that quite probably it will forever remain incomplete. One can only make so many silk purses out of pig's ears; less metaphorically, one can only discern so much about the past in the light of the missing and presumably irretrievably lost information. The simple fact of the matter is that most organisms do not get fossilized and that there are altogether too many intervening factors to compensate for this loss through other techniques, such as those used in molecular biology. So whereas the fact of evolution is surely now a well-established proposition, paths of evolution will always be at some level incomplete, if only in practice, if not in theory, beyond the limits of our understanding.

## THE CAUSES OF EVOLUTION

In the *Origin*, Darwin argued for the fact of evolution. He said little about paths, except in a general sense about progress and branching. But he did argue strenuously for a mechanism: natural selection. Starting from the Malthusian struggle for existence, Darwin argued that given a constant supply of new heritable variation there will be an inevitable ongoing process akin to the artificial selection practiced by animal and plant breeders. This natural selection, given enough time, leads to full-blown evolution. Not mere evolution either, but the evolution of well-adapted, or functioning, organisms; living begins with characteristics or "adaptations" directed toward their possessors' well-being.

Although people quickly accepted the fact of evolution, acceptance of Darwin's mechanism of natural selection—and a secondary mechanism, sexual selection, based on a struggle for mates—was much longer in coming. Indeed, for the first 75 years of post-*Origin* evolutionism, natural selection was relegated to a very minor role (Ruse, 1979a, 1996; Bowler, 1983). However, by the 1930s with the discovery and articulation of Mendelian genetics, it was realized that the underpinning stratum for natural selection was now in place, and people swung very rapidly to a synthesis of

Darwinian selection and Mendelian genetics (Provine, 1971). A synthesis often known as "Neo-Darwinism." By the 1950s and the hundredth anniversary of the *Origin*, this theory reigned supreme, especially in English-speaking countries (Tax, 1960a,b; Tax and Calender, 1960). It seemed to those at the forefront of evolutionary biology of that day that the main outlines of an entirely satisfactory causal theory of evolution was now at hand. All that was necessary was the filling in of the details.

Viewed from some 40 years on, that estimation seems to many to be very optimistic, unreasonably so. This is certainly not because selection has failed as mechanism, collapsing in on itself. Indeed, in many respects it has triumphed in the second century of its history as never before. One thinks particularly of the work of the students of social behavior, the so-called "sociobiologists" (Ruse, 1979b; Wilson, 1971, 1975). Thanks to a number of models applied to social behavior—the kin selection proposals of William Hamilton (1964a,b), the evolutionarily stable strategies of John Maynard Smith (1978a, 1982), and more (Trivers, 1971)—we have a selection-based understanding of animal social behavior as could only have been dreamed of by Darwin and his contemporaries. Moreover, we find the newest of biological theories coming to the aid of selection studies; a prime example is the way molecular biology swings into action, especially through the use of so-called genetic fingerprinting in order to trace paternities, thus providing confirmation of predictions arrived at by sociobiological theory and observation (Davies, 1992). It would be a rash person indeed who would predict that in the area of social behavior there are any limits to our possible understanding of the evolutionary process.

But staying for the moment at the whole-organism level of inquiry, one cannot pretend that all is entirely smooth and set for untroubled continued success. There are, as we now realize, still some outstanding and fundamental theoretical questions about the ways in which selection at a general level operates in populations. Common opinion, in North America particularly, was that the population geneticist Sewall Wright had offered key insights at the beginning of the 1930s with his "shifting balance theory of evolution," a picture that incorporates the metaphor of an adaptive landscape, with populations being dislodged from peaks of fitness by random processes (notably "genetic drift") and then scaling yet greater heights under the influence of selection (Wright, 1986a,b). Recently, however, this theory has been subjected to withering criticism, and little remains but tattered fragments (Coyne *et al.*, 1997).

It is true that the authors of the critique take this as a reason to revert to an even more selection-based position than before; but dare one predict that selection will prove to have the capacity to solve all the major outstanding problems? A running sore is that of the origination and maintenance of sexuality. In the old days, people tended to think of selection working at different levels, including that of the group; hence, sexuality could be explained simply in terms of the increased benefits to the species, brought about by combining rare but favorable mutations in a single organism through the sexual process. Now, however, evolutionists tend to veer away from group selection explanations, except in very special cases, and it is difficult to see how sexuality can be maintained in the light of pressures always favoring the individual (Williams, 1975; Maynard Smith, 1978b). It is true that there are models (using individual selection) that have been proposed to explain sexuality. For instance, Hamilton (Hamilton *et al.*, 1990) has suggested that it exists in order to provide protection against parasites (basically the idea being that sexuality shuffles the genes, and thus presents parasites with an ever-moving target). Others have suggested mechanisms more at the molecular level (Michod and Levin, 1988). Yet while many of these mechanisms or models have merit, not one at this point commands universal attention and acceptance. [A recent special issue of *Science* deals with many of the issues around the evolution of sex. See especially Wuethrich (1998), Barton and Charlesworth (1998), and Ryan (1998).]

None of this of course suggests that the problem of sexuality and like problems are insoluble; I doubt that any of the researchers on the problem of the evolution of sexuality would want to claim that there are limits to our understanding here or that these limits have been reached or will shortly be reached. It is one thing not to have answers to all of the problems; it is another to say that we can and will never have answers to all of the problems. But one should be aware that there is optimism here not always matched by results. Especially in the light of the collapse of the hitherto universally feted shifting balance theory, it would be naive to presume either that already existing alternatives (Fisher, 1930) will be found fully satisfactory or that a new universal theory will soon take its place. Right at its heart, there are serious questions about contemporary evolutionary theory and about our abilities to resolve the issues.

Doubts at the whole-organism level are matched by doubts at the molecular level. Again, let us recognize that in the past 40 years there have been major advances, notably in the understanding of the genetic compositions of populations. This progress is due particularly to breakthrough techniques (gel electrophoresis) pioneered by Richard Lewontin and others, showing how molecular genetic variation can be documented in virtually every species and how this variation can be related to evolutionary processes (Hubby and Lewontin, 1966; Lewontin and Hubby, 1966). However, again we start to encounter doubts and worries about our prospects of ever achieving full understanding of evolution at the causal level. There is a sense that, almost paradoxically, the great successes backfire, inasmuch as they show that the problems remaining are greater yet than ever conceived. Lewontin particularly was and remains deeply pessimistic about the possibility of a full understanding of the processes of evolution at the genetic level, and indeed feels that the increased information achieved by his own breakthrough techniques if anything make the task more formidable rather than less (Lewontin, 1974, 1991a). He argues that the theories that we have at the moment are simply not capable of leading to a full understanding of the nature of biological variation. He concludes that the prospects, at least at present, for improved theories better able to handle the problems are slight indeed. I am sure that many do not share this deep pessimism; but undoubtedly if not theoretical limits there do seem to be major practical limits to our understanding of molecular biological variation. Compounding this are the difficulties that are raised when one moves further into such complicating questions as development. Here again, particularly at the molecular level, major moves forward are being made, but there remains much work to be done, and it would be an optimist indeed who would predict confidently that this work will soon come to a complete and satisfactory end.

## MACROEVOLUTION

Much of this is a question of perspective. Whether or not there are deep theoretical and practical problems, there is much exciting work to be done and not just in the social realm. Many evolutionists feel that not only are evolutionary studies at the causal level vigorous, but that they are moving forward in a satisfactory manner. Any talk on limits to our possible understanding will be greeted with some skepticism, if not scorn. To use the language of the philosopher-historian Thomas Kuhn (1962), evolutionists have got a paradigm and they are not inclined to think about its limits. Indeed, contemplating limits is self-defeating, for then one will surely find them. But continuing to play the skeptic, let me note that most professional evolutionary theorizing centers on microstudies of the evolutionary process. As Stephen Jay Gould (1980, 1982, 1989) and others have long cautioned, when one moves into the macro realm, things start to look a little different. One should not be unduly pessimistic about this. Major moves forward have occurred with respect to understanding evolution at the macrolevel. One particularly promising avenue of research has been that provided by the American paleontologist John Sepkoski (1978, 1979, 1984), who has applied the MacArthur–Wilson (1967) theory of island biogeography to temporal problems, showing how one can readily explain such large-scale phenomena as the huge rise in the numbers of species and higher taxa during the Cambrian period (500–600 million years ago). Sepkoski shows that the sigmoidal-type growth one finds in the that period is precisely that which one would expect were nature obeying the same principles that have been worked out at the ecological level dealing with the invasion of space (Figs. 8 and 9). Moreover, Sepkoski is able to show in refinements of his theorizing how the MacArthur–Wilson island biogeography theory accounts for subsequent decline of Cambrian-type organisms and the rise of later forms. Indeed, he can even apply this type of reasoning to more specific phenomena such as the joining of the North and South Americas, and the subsequent invasion of the North by southern forms and the South by northern forms, with numbers of taxa peaking and then declining from their optimum (Marshall *et al.*, 1982).

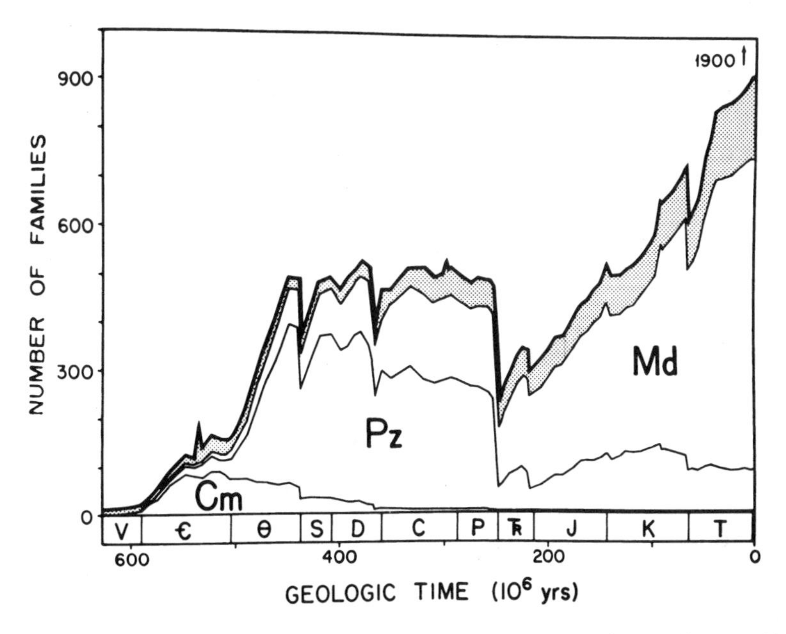


FIG. 8. The history of life since the Cambrian explosion. (From Sepkoski, 1984. Used with permission of the author.)

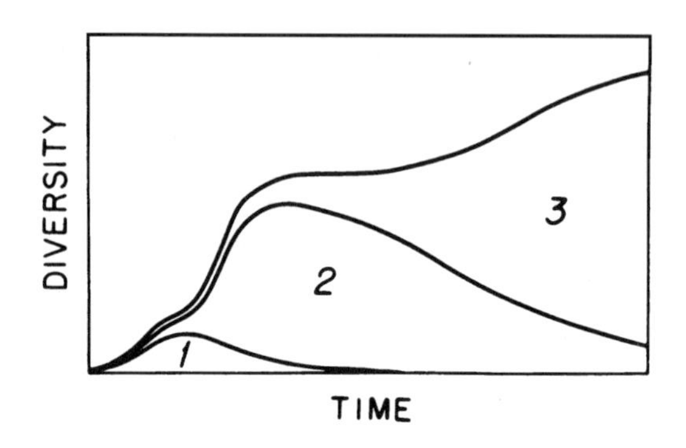


FIG. 9. Models of life's history. (From Sepkoski, 1984. Used with permission of the author.)

Yet for all this success, we are far from a complete understanding of the processes of life's macrohistory. At a minimum, there are complicating factors like random external (that is, nonbiological) events, such as massive extinctions possibly caused by extraterrestrial factors like earth-impacting asteroids (as surely occurred at the time when the dinosaurs vanished from the earth) (Raup, 1986; Raup and Sepkoski, 1982; Alvarez *et al.*, 1984). At a maximum, there are questions about whether even in principle orthodox selection-based Darwinian theory can do what is needed. Gould promotes his own alternative of punctuated equilibria, supposing sharp breaks in the course of evolution and downplaying the significance of selection in favor of nonadaptive "constraints" on development (Eldredge and Gould, 1972; Gould and Eldridge, 1977, 1993; Gould and Vrba, 1982; Gould, 1992). I am not sure that this proposal is judged a great success, but whatever the fate of that one particular theory, few would claim that no such theory will ever prove useful or needed or that it will be discovered, for that matter. Perhaps, as with questions about the actual paths of evolution, discussions of the causes of evolution over vast periods of life history are doomed for ever to be incomplete.

## HUMAN EVOLUTION

Finally, completing this litany of doubt, let me turn to what will strike many as the most likely place where we will encounter limits to our knowledge of the causes of evolution. I refer to the human realm. No one today wants to deny that humans are the product of evolution, or even that natural selection played a major rule in human evolution. But it hardly needs saying that there are many today, even within the biological community, who are extremely dubious about the extent to which evolutionary biology can throw meaningful light on human nature

(Gould and Lewontin, 1979; Gould, 1981; Kitcher, 1985; Lewontin, 1991b; Lewontin *et al.*, 1984). They believe that, whatever our past may have been, humans today have entered the whole new dimension of culture, a dimension in which biological factors play little or no role. In this respect, therefore, there are and always will be major limits to our knowledge of evolution; limits because of the complexity that culture brings with it, as well as limits imposed by the constraints of culture itself. Even if Nazi-type experiments could yield uniquely true insights, no one would want to pay the price.

One trusts that no one denies the moral limits to our knowledge, certainly not me. But, let me say directly that I am not one who believes that culture is an insuperable barrier to any meaningful understanding of human (causal) evolution. I prefer to see biology and culture working in tandem (Ruse, 1986); but if they are to be considered apart, if only for conceptual reasons, I assert that biology is if anything the most powerful tool that we have for the understanding of human nature. Not only are we the product of evolution, but in order to understand us as we function today one must understand our past and the evolutionary processes still operating (Pinker, 1997; Deacon, 1997; Betzig *et al.*, 1987). Foremost among these has been and in many respects continues to be natural selection. I would argue that the human sociobiologists or, as they are sometimes now referred to, the human behavioral ecologists or evolutionary psychologists, have made major inroads into an understanding of human nature: in the ways that we reason, in the ways that we interact socially, and much more. I consider it beyond reasonable doubt that relationships between the sexes or between the generations in many respects reflect our evolutionary heritage. Studies on homicide, for instance, are outstanding, showing how those who commit such acts are precisely those whom evolution identifies as the likely culprits (Daly and Wilson, 1988).

However, having thus endorsed the causal success of human evolutionary biology and having expressed confidence in its future, I admit that it would be naive indeed to suggest that we are close to a full understanding of human nature from an evolutionary perspective, or even that we will ever actually complete a full understanding. Human nature is so diverse, both within and between societies and especially across such short periods of time, that hopes of achieving an evolutionary science of human nature of a kind or degree that we find (say) in the physical sciences is not simply a vain hope but probably a self-defeating hope. Nothing in the history of science suggests otherwise. To try to achieve in human evolutionary biology such predictive accuracy as we find in physics probably demands such a constriction or narrowing of focus in the very effort that one necessarily will ignore the full experiential richness of the human condition.

At the very least, therefore, one must leave open the question of whether the full science of human nature from an evolutionary perspective will ever be achieved, or is even in principle achievable. This is not so much because humans are in principle different from other organisms, but rather because the complexity one faces when dealing with humans, a complexity that is ever changing at rates virtually unknown elsewhere in the organic world, makes any hopes of a fully comprehensive understanding almost self-defeating.

## ALTERNATIVE THEORIES?

I have argued that although the fact of evolution is well established, there are probably theoretical and almost certainly practical reasons why we will never fully grasp the paths or the causes of the evolutionary process. I do not find this conclusion particularly radical or surprising: It is surely one that would be agreed on by most evolutionists themselves, even though they might perhaps draw lines in different places from those that I have drawn or indeed from those that their fellow evolutionists might draw. However, before concluding it is only fair and proper to consider some of the more radical attacks that one might make on this particular topic. These attacks might come from two sources: scientific and philosophical. First, scientific, from those who would argue that we already see new dimensions to evolutionary causal understanding, and that when once these dimensions are considered, perhaps the question of limits ought to be reconsidered. Then second, philosophical, from those who would attack the very idea of science as it has been conceived by most people since the Enlightenment. They too might think that the limits question ought to be reconsidered. Let me take these two attacks, briefly, in turn.

My first group of thinkers are those who, once again to use Kuhnian language, are challenging the dominant paradigm. These are the people who are deeply critical of the view that evolution is a process fueled essentially by natural selection. In this group one might possibly find the already-mentioned Stephen Jay Gould and his theory of punctuated equilibria; although, to be quite frank, given the ambiguity and many formulations of this theory, it is difficult to know whether it is being offered as a new paradigm (Ruse, 1989, 1994, 1999). One would certainly find those who think that nonequilibrium thermodynamical systems hold the clues to life's history (Brooks and Wiley, 1986; Depew and Weber, 1994). Those who believe that organic complexity is the reflection of some deep underlying physical structures, more to do with the very nature of matter and less specifically as a function of the purely biological also are included. One has in mind here people like Stuart Kauffman (1993) and others at the Santa Fe Institute, who argue—in a way reminiscent of the early 19th-century German morphological-*cum*-philosophical movement known as *Naturphilosophie*—that nature itself incorporates certain principles of order, and that these manifest themselves in organisms (Lewin, 1992). Organisms are as they are, not because of their biological fitness, but because these are the fundamental ways in which matter organizes itself (Gould, 1977).

I should say straight off that I myself am not particularly sympathetic to these alternatives. Whatever its troubles or limits, I fail to entirely see that Darwinism is better replaced by radically new alternatives, be these extrapolations from contemporary physics or reversions to 19th-century German philosophy. But here my own personal predilections are beside the point. What I would readily agree is that if one or more of these alternatives does indeed prove in the future to be more fruitful than the essentially dominant Darwinian position, then obviously one is going to rethink the question of the limits to knowledge. Just to take one example. As I have pointed out, when thinking of things at the macrolevel, the Darwinian does rather come up against the problem of random external forces. However powerful a mechanism natural selection truly proves to be, it is not the only causal factor affecting life's history. For instance, there may be repeated extraterrestrial interventions—falling asteroids and the like. These obviously cannot be readily quantified and incorporated within a biological theory; they will probably always generate a level of randomness, putting limits on our possible scientific understanding. Selection is a mechanism that changes as the circumstances change, and if the circumstantial changes are outside the domain of understanding, so also in some sense are the selective responses. However, if one embraces one of the alternative theories, one might then find that life's patterns occur irrespective of external factors, being perhaps part of the built-in nature of the organic process. If this were the case, one would be able to extend or transcend limits that are set by contemporary evolutionary biology.

Analogously, if one had some alternative theory of human nature—evolutionary but not strictly Darwinian—one might again have a picture where hitherto nontransgressible limits to knowledge are now places where we can move forward with optimism and hope. Note, however, the Kuhnian point that when one has new paradigms, the opening of new vistas is often accompanied by the setting of restrictions on what hitherto had been considered fruitful. Under these new paradigms, areas where Darwinism thinks that it has possibly pushed the limits of ignorance back quite considerably would now no longer be considered particularly interesting or significant at all. New paradigms would change the limits of knowledge, but one should not assume automatically that these limits would all be further from where we are now.

The second, more philosophical attack starts with the fact that many commentators on science today argue that science is less a disinterested objective reflection of a separate mind-independent reality and more a subjective human-created "social construction." In many respects science is merely an epiphenomenon of culture, something that changes less at the dictates of a mind-independent reality (which may not exist anyway) and more because of the changes of culture itself (Collins, 1985; Latour, 1987; Latour and Woolgar, 1979). If this is the case, then one ought not privilege contemporary evolutionary theory or even any of its alternatives. One ought not argue that Darwinism or neo-Darwinism or one of the alternatives is uniquely and forever the true or right way to understand life's history. Rather, one must recognize that such understanding will always be culture dependent and forever subject to revision and change in the light of cultural change. And obviously here, even more than before, one has the prospect of radical revision of our thinking about possible limits to understanding. As culture changes so also will the evolutionary theories; as these theories change, so also will the prospects of limits.

Let me say that, committed Darwinian though I may be, I am not entirely unsympathetic to this line of argument. One cannot take the history of science seriously without appreciating that scientific theories, including Darwinian evolutionary theory, are deeply imbedded in their culture, and that as the culture changes so also does the science (Ruse, 1999). In the case of Darwinism, only the ignorant could deny that it is deeply rooted in the mentality of the 18th-century Enlightenment: It would not exist save for the political economy of thinkers like Adam Smith; the progressionism of idealists like Condorcet; not to mention the simple and straightforward agricultural advances so cherished by the early evolutionist and grandfather of Charles Darwin, Erasmus Darwin, and his circle in the Lunar Society (Ruse, 1996). Nor should one pretend that today's evolutionary biology escapes its culture: Think of the extent to which today's sociobiologists, happily seeking evolutionarily stable strategies, rely so heavily on game theory, a branch of mathematics developed in response to needs by the military and industry during and after World War II. If anything, with the talk of "strategies" and related concepts like "arms races" and "selfish genes" and the like, evolutionary biology is more imbedded in its culture today than it was back in the 18th century (Dawkins, 1986; Dawkins and Krebs, 1979). Hence, it is plausible to argue that were society to change, evolutionary biology would change, and that had society been different from that which it is, our understanding would not be as it is.

This being the case, the whole question of limits becomes a much more subjective and culturally infected matter than it would appear under a more traditional, objectivist philosophy of science. [The kind of philosophy associated with the late Sir Karl Popper (1972).] However, having said this much, I do not intend my conclusions to be quite as subversive as might seem. Were society different, radically so, then it is quite possible that our thinking would be different, radically so. But it does not follow that we would then simply have alternative evolutionary theories, such as punctuated equilibria, to conventional Darwinism. We might. There is good evidence that late 19th-century Russian evolutionism was of the same logical type as Western European evolutionism but differed from it because of cultural factors: the Russians could not envision a situation where the struggle for existence was more intense between individuals than with the elements, and hence they were led to theories of evolutionary cooperation (Kropotkin, 1955; Todes, 1989). But given the alternative culture, one might have a situation where one was not thinking in a way in which one would want to call evolutionary at all. In which case, the entire question of limits to evolutionary understanding is rather less dramatically affected than might seem the case at first sight.

Let me spell out my point. Although our understanding of evolution is surely and deeply something that is part of Western post-Renaissance thinking, the roots go back much farther. Indeed, the whole practice of asking about origins—and demanding historical answers—is something that goes back to pre-Christian Jewish thought (and mid-Eastern thinking generally). Questions of origins did not necessarily take on the same form or vigor in other cultures (Ruse, 2000). The ancient Greeks, for instance, although they were interested in origins, did not provide the kind of historical account that one finds in Genesis. Plato thought in terms of cycles and Aristotle in terms of eternity. My point therefore is that in a different society we might not ask questions about origins in the way that we do. Indeed, origins might simply not be a very interesting question at all. This being so, we might not provide answers in the way that we do, referring to such 18th-century political economic ideas as the struggle for existence, or 19th-century ideas about natural selection, or 20th-century ideas about strategies and arms races. Therefore, not only might we not have an evolutionary theory of the kind that we do have, we might not have an evolutionary theory at all—and not miss it. We would not be asking questions that demanded the kinds of answers that evolution is inclined to offer.

So I suggest that the consequence of the social constructivist argument is perhaps even more radical, and, paradoxically, less significant than the constructivists themselves realize. It is not that a different culture would simply lead to a different evolutionary theory, but that a different culture might lead to no evolutionary theory at all. I am not now saying that such a science would be impoverished or that evolution as we understand it would no longer be there for anyone to discover, but rather that in such a culture we would be dividing up the empirical pie in completely different ways. Interest in origins would not necessarily be considered significant, whereas other questions would loom larger than they do now. What these other questions would be I am afraid I am not really able to answer, because I am not part of such a society, but it seems plausible to me that they might exist. Whatever the case may be, worries about the limits to our understanding of evolution would simply not arise, because evolution would not be a matter of concern or interest.

My conclusion therefore is that the constructivist argument is perhaps so powerful as to render otiose or nonthreatening the very concerns it seems initially to raise. But I suspect that this is a conclusion that excites me more than others. I fear that by this stage I have become so philosophical and beyond the realm of the actual that I have lost most of my readers, certainly most of my scientific readers. Therefore this is a good point at which to bring my discussion to an end. We know much about evolution. The fact of evolution is firmly established. We know much about the paths of evolution and we know much about the causes or mechanisms of evolution. However, one has to admit that our knowledge of paths and of causes is limited and probably always will be limited. Research moves forward today perhaps more vigorously and in a more exciting fashion that ever before, but it would be naive and wrong to pretend that there are no limits. What is important is to find a balance; not to let the thought of limits stifle activity, but not to so ignore the possible limits to our understanding that we become narrow and crude and unimaginative in a false attempt to break through to unscaled heights.

I conclude with one of my favorite reflections by the population geneticist J. B. S. Haldane (1927): "My own suspicion is that not only is the universe queerer than we suppose, but queerer than we *can* suppose" (p. 298). We have no reason to think that middle-range primates, using the adaptations they evolved to come down from out of the jungle, necessarily have the powers to achieve full understanding of their origins. The wonder is that we can know as much as we do.

## REFERENCES

Allen, G. E., 1978, *Life Science in the Twentieth Century*, Cambridge University Press, Cambridge, England.

Alvarez, W., Kauffmann, E. G., Surlyk, R., Alvarez, L. W., Asaro, R., and Michel, H. V., 1984, Impact theory of mass extinctions and the invertebrate fossil record, *Science* **223:**1135–1141.

Ayala, F. J., 1978, The mechanisms of evolution, *Sci. Am.* **Sept:**56–69.

Ayala, F. J., 1985, The theory of evolution: Recent successes and challenges, in: *Evolution and Creation* (E. McMullin, ed.), pp. 59–90, University of Notre Dame Press, Notre Dame, Indiana.

Bartel, D. P., and Szostak, J. W., 1993, Isolation of new ribozymes from a large pool of random sequences, *Science* **261:**1411–1418.

Barton, N. H., and Charlesworth, B., 1998, Why sex and recombination? *Science* **281:**1986–1990.

Behe, M., 1996, *Darwin's Black Box: The Biochemical Challenge to Evolution*, Free Press, New York.

Betzig, C. C., Borgerhoff Mulder, M., and Turke, P. W., 1987, *Human Reproductive Behavior*, Cambridge University Press, Cambridge, England.

Bowler, P. J., 1976, *Fossils and Progress*, Science History Publications, New York.

Bowler, P. J., 1983, *The Eclipse of Darwinism: Anti-Darwinism Evolution Theories in the Decades around 1900*, Johns Hopkins University Press, Baltimore.

Brooks, D., and Wiley, E. O., 1986, *Evolution as Entropy: Toward a Unified Theory*, University of Chicago Press, Chicago.

Cairns-Smith, A. G., 1982, *Genetic Takeover and the Mineral Origins of Life*, Cambridge University Press, Cambridge, Massachusetts.

Cairns-Smith, A. G., 1986, *Clay Minerals and the Origin of Life*, Cambridge University Press, Cambridge, Massachusetts.

Carroll, R. L., 1997, *Patterns and Processes of Vertebrate Evolution*, Cambridge University Press, Cambridge, Massachusetts.

Coleman, W., 1964, *Georges Cuvier Zoologist. A Study in the History of Evolution Theory*, Harvard University Press, Cambridge, Massachusetts.

Collins, H., 1985, *Changing Order*, Sage, London.

Coyne, J. A., Barton, N. H., and Turelli, M., 1997, Perspective: A critique of Sewall Wright's shifting balance theory of evolution, *Evolution* **51:**643–671.

Crick, F. H. C., and Orgel, L. E., 1973, Directed panspermia, *Icarus* **19:**341–346.

Daly, M., and Wilson, M., 1988, *Homicide*, De Gruyter, New York.

Darwin, C., 1859, *The Origin of Species*, John Murray, London.

Darwin, C., 1871, *The Descent of Man*, John Murray, London.

Davies, N. B., 1992, *Dunnock Behaviour and Social Evolution*, Oxford University Press, Oxford, England.

Dawkins, R., 1986, *The Blind Watchmaker*, Norton, New York.

Dawkins, R., and Krebs, J. R., 1979, Arms races between and within species, *Proc. Roy. Soc. Lond. B* **205:**489–511.

Deacon, T. W., 1997, *The Symbolic Species: The Co-Evolution of Language and the Brain*, Norton, New York.

Depew, D., and Weber, B., 1994, *Darwinism Evolving*, MIT Press, Cambridge, Massachusetts.

Dobzhansky, T., Ayala, F. J., Stebbins, G. L., and Valentine, L., 1977, *Evolution*, W. H. Freeman, San Francisco.

Doolittle, R. F., 1997, A delicate balance, *Boston Rev.* **22(1):**28–29.

Ekland, E. H., and Bartel, D. P., 1996, RNA-catalysed RNA polymerization using nucleotide triphosphates, *Nature* **382:**373–376.

Ekland, E. H., Szostak, J. W., and Bartel, D. P., 1995, Structurally complex and highly active RNA ligases derived from random RNA sequences, *Science* **269:**364–370.

Eldredge, N., and Gould, S. J., 1972, Punctuated equilibria: An alternative to phyletic gradualism, in: *Models in Paleobiology* (T. J. M. Schopf, ed.), pp. 82–115, Freeman, Cooper, San Francisco.

Feduccia, A., 1996, *The Origin and Evolution of Birds*, Yale University Press, New York.

Ferris, J. R., Hill, A. R., Liu, R., and Orgel, L. E., 1996, Synthesis of long prebiotic oligomers on mineral surfaces, *Nature* **381:**59–61.

Fisher, R. A., 1930, *The Genetical Theory of Natural Selection*, Oxford University Press, Oxford, England.

Fox, S. W., 1988, *The Emergence of Life: Darwinian Evolution from the Inside,* Basic Books, New York.

Freeman, S., and Herron, J. C., 1998, *Evolutionary Analysis*, Prentice-Hall, Englewood-Cliffs, New Jersey.

Gould, S. J., 1977, *Ontogeny and Phylogeny*, Belknap Press, Cambridge, Massachusetts.

Gould, S. J., 1980, Is a new and general theory of evolution emerging? *Paleobiology* **6:**119–130.

Gould, S. J., 1981, *The Mismeasure of Man*, Norton, New York.

Gould, S. J., 1982, Darwinism and the expansion of evolutionary theory, *Science* **216:**380–387.

Gould, S. J., 1989, *Wonderful Life: The Burgess Shale and the Nature of History*, Norton, New York.

Gould, S. J., 1992, Punctuated equilibrium in fact and theory, in: *The Dynamics of Evolution: The Punctuated Equilibrium Debate in the Natural and Social Sciences*, (A. Somit and S. A. Peterson, eds.) pp. 54–84, Cornell University Press, Ithaca, New York.

Gould, S. J., and Eldredge, N., 1977, Punctuated equilibria: the tempo and mode of evolution reconsidered, *Paleobiology* **3:**115–151.

Gould, S. J., and Eldredge, N., 1993, Punctuated equilibrium comes of age, *Nature* **366:**223–227.

Gould, S. J., and Lewontin, R. C., 1979, The spandrels of San Marco and the Panglossian paradigm: A critique of the adaptationist program, *Proc. Roy. Soc. Lond. B* **205:**581–598.

Gould, S. J., and Vrba, E. S., 1982, Exaptation—A missing term in the science of form, *Paleobiology* **8:**4–15.

Haldane, J. B. S., 1927, *Possible Worlds, and Other Papers*, Chatto and Windus, London.

Hamilton, W. D., 1964a, The genetical evolution of social behaviour I, *J. Theor. Biol.* **7:**1–16.

Hamilton, W. D., 1964b, The genetical evolution of social behaviour II, *J. Theor. Biol.* **7:**17–32.

Hamilton, W. D., Axelrod, R., and Tanese, R., 1990, Sexual reproduction as an adaptation to resist parasites, *Proc. Nat. Acad. Sci. USA* **87:**3566–3573.

Hubby, J. L., and Lewontin, R. C., 1966, A molecular approach to the study of genic heterozygosity in natural populations. I. The number of alleles at different loci in *Drosophila pseudoobscura*, *Genetics* **54:**577–594.

Johanson, D., and Edey, M., 1981, *Lucy: The Beginnings of Humankind*, Simon and Schuster, New York.

Johnson, P. E., 1991, *Darwin on Trial*, Regnery Gateway, Washington, DC.

Johnson, P. E., 1995, *Reason in the Balance: The Case against Naturalism in Science, Law and Education*, InterVarsity Press, Downers Grove, Illinois.

Kauffman, S. A., 1993, *The Origins of Order: Self-Organization and Selection in Evolution*, Oxford University Press, Oxford, England.

Kitcher, P., 1985, *Vaulting Ambition*, MIT Press, Cambridge, Massachusetts.

Kropotkin, P., 1955, *Mutual Aid*, Extending Horizons Books, Boston. (Originally Published 1902.)

Kuhn, T., 1962, *The Structure of Scientific Revolutions*, University of Chicago Press, Chicago.

Latour, B., 1987, *Science in Action*, Harvard University Press, Cambridge, Massachusetts.

Latour, B., and Woolgar, S., 1979, *Laboratory Life: The Construction of Scientific Facts*, Sage, Beverly Hills, California.

Lewin, R., 1992, *Complexity: Life at the Edge of Chaos*, Collier, New York.
Lewontin, R. C., 1974, *The Genetic Basis of Evolutionary Change* (XXV in the Columbia Biology Series), Columbia University Press, New York.
Lewontin, R. C., 1991a, 25 years ago in genetics—Electrophoresis in the development of evolutionary genetics—milestone or millstone? *Genetics* **128:**657–662.
Lewontin, R. C., 1991b, *Biology as Ideology. The Doctrine of DNA*, Anansi, Toronto.
Lewontin, R. C., and Hubby, J. L., 1966, A molecular approach to the study of genic heterozygosity in natural populations. II. Amount of variation and degree of heterozygosity in natural populations of *Drosophila pseudoobscura*, *Genetics* **54:**595–609.
Lewontin, R. C., Rose, S., and Kamin, L. J., 1984, *Not in Our Genes*, Pantheon, New York.
MacArthur, R. H., and Wilson, E. O., 1967, *The Theory of Island Biogeography*, Princeton University Press, Princeton, New Jersey.
Maienschein, J., 1991, *Transforming Traditions in American Biology 1880–1915*, Johns Hopkins University Press, Baltimore.
Marshall, L. G., Webb, S. D., Sepkoski, Jr., J. J., and Raup, D. M., 1982, Mammalian evolution and the great American interchange, *Science* **215:**1351–1357.
Maynard Smith, J., 1978a, The evolution of behavior, *Sci. Am.* **239(3):**176–193.
Maynard Smith, J., 1978b, *The Evolution of Sex*, Cambridge University Press, Cambridge, England.
Maynard Smith, J., 1982, *Evolution and the Theory of Games*, Cambridge University Press, Cambridge, England.
Michod, R. E., and Levin, B. R. (eds.), 1988, *The Evolution of Sex*, Sinauer, Sunderland, Massachusetts.
Miller, S. L., 1953, A production of amino acids under possible primitive earth conditions, *Science* **117:**528–529.
Miller, S. L., 1992, The prebiotic synthesis of organic compounds as a step toward the origin of life, in: *Major Events in the History of Life* (J. W. Schopf, ed.), pp. 1–28, Jones and Barlett, Boston.
Ostrom, J. H., 1976, Archaeopteryx and the origin of birds, *Biol. J. Linn. Soc.* **8:**91–182.
Owen, R., 1861, *Paleontology*, 2nd ed., Black, Edinburgh.
Pinker, S., 1997, *How the Mind Works*, Norton, New York.
Plantinga, A., 1991a, When faith and reason clash: Evolution and the Bible, *Christ. Schol. Rev.* **21:**8–32.
Plantinga, A., 1991b, Evolution, neutrality, and antecedent probability: A reply to Van Till and McMullin, *Christ. Schol. Rev.* **21:**80–109.
Popper, K. R., 1972, *Objective Knowledge*, Oxford University Press, Oxford, England.
Provine, W. B., 1971, *The Origins of Theoretical Population Genetics*, University of Chicago Press, Chicago.
Rainger, R., 1991, *An Agenda for Antiquity: Henry Fairfield Osborn and Vertebrate Paleontology at the American Museum of Natural History*, 1890–1935, University of Alabama Press, Tuscaloosa.
Raup, D. M., 1986, *The Nemesis Affair: A Story of the Death of Dinosaurs and the Ways of Biology*, Norton, New York.
Raup, D. M., and Sepkoski, Jr., J. J., 1982, Mass extinctions in the marine fossil record, *Science* **215:**1501–1503.
Richards, R. J., 1992, *The Meaning of Evolution: The Morphological Construction and Ideological Reconstruction of Darwin's Theory*, University of Chicago Press, Chicago.
Ridley, M., 1986, *Evolution and Classification: The Reformation of Cladism*, Longman, New York.
Ridley, M., 1993, *Evolution*, Blackwell, Boston.

Rudwick, M. J. S., 1972, *The Meaning of Fossils*, Science History Publications, New York.
Ruse, M., 1973, *The Philosophy of Biology*, Hutchinson, London.
Ruse, M., 1979a, *The Darwinian Revolution: Science Red in Tooth and Claw*, University of Chicago Press, Chicago.
Ruse, M., 1979b, *Sociobiology: Sense or Nonsense?* Reidel, Dordrecht, The Netherlands.
Ruse, M., 1982, *Darwinism Defended: A Guide to the Evolution Controversies*, Addison-Wesley, Reading, Massachusetts.
Ruse, M., 1984, Is there a limit to our knowledge of evolution? *BioScience* **34:**100–104.
Ruse, M., 1986, *Taking Darwin Seriously*, Blackwell, Oxford, England.
Ruse, M., 1988a, *But Is It Science? The Philosophical Question in the Creation/Evolution Controversy*, Prometheus, Buffalo, New York.
Ruse, M., 1988b, *Philosophy of Biology Today*, State University of New York Press, Albany.
Ruse, M., 1989, *The Darwinian Paradigm: Essays on its History, Philosophy and Religious Implications*, Routledge, London.
Ruse, M., 1994, *Evolutionary Naturalism: Selected Essays*, Routledge, London.
Ruse, M., 1996, *Monad to Man: The Concept of Progress in Evolutionary Biology*, Harvard University Press, Cambridge, Massachusetts.
Ruse, M., 1997, The origin of life: philosophical perspectives, *J. Theor. Biol.* **187:**473–482.
Ruse, M., 1999, *Mystery of Mysteries: Is Evolution a Social Construction?* Harvard University Press, Cambridge, Massachusetts.
Ruse, M., 2000, *Can a Darwinian Be a Christian? One Person's Answer*, Cambridge University Press, Cambridge, Massachusetts (in press).
Ryan, M. J., 1998, Sexual selection, receiver biases, and the evolution of sex differences, *Science* **281:**1999–2003.
Scott, E. C., 1997, Antievolutionism and creationism in the United States, *Annu. Rev. Anthropol.* **26:**263–289.
Sepkoski, Jr., J. J., 1978, A kinetic model of Phanerozoic taxonomic diversity. I Analysis of marine orders, *Paleobiology* **4:**223–251.
Sepkoski, Jr., J. J., 1979, A kinetic model of Phanerozoic taxonomic diversity. II Early Paleozoic families and multiple equilibria, *Paleobiology* **5:**222–252.
Sepkoski, Jr., J. J., 1984, A kinetic model of Phanerozoic taxonomic diversity. III Post-Paleozoic families and mass extinctions, *Paleobiology* **10:**246–267.
Sober, E., 1988, *Reconstructing the Past: Parsimony, Evolution, and Inference*, MIT Press, Cambridge, Massachusetts.
Strickberger, M., 1990, *Evolution*, Jones and Bartlett, Boston.
Tax, S. (ed.), 1960a, *Evolution After Darwin. I The Evolution of Life*, Chicago University Press, Chicago.
Tax, S. (ed.), 1960b, *Evolution after Darwin. II The Evolution of Man*, Chicago University Press, Chicago.
Tax, S., and Calender, C. (eds.), 1960, *Evolution After Darwin. III Issues in Evolution*, Chicago University Press, Chicago.
Todes, D. P., 1989, *Darwin without Malthus: The Struggle for Existence in Russian Evolutionary Thought*, Oxford University Press, New York.
Trivers, R., 1971, The evolution of reciprocal altruism, *Q. Rev. Biol.* **46:**35–57.
Valentine, J. W., 1978, The evolution of multicellular plants and animals, *Sci. Am.* **8:**140–158.
Whewell, W., 1840, *The Philosophy of the Inductive Sciences*, (2 vols.), Parker, London.
Williams, G. C., 1975, *Sex and Evolution*, Princeton University Press, Princeton, New Jersey.
Wilson, E. O., 1971, *The Insect Societies*, Harvard University Press, Cambridge, Massachusetts.
Wilson, E. O., 1975, *Sociobiology: The New Synthesis*, Harvard University Press, Cambridge, Massachusetts.

Wilson, E. O., 1998, *Consilience*, Knopf, New York.
Wright, S., 1986a, Evolution in Mendelian populations, in: *Evolution: Selected Papers* (W. B. Provine, ed.), Chicago University Press, Chicago. (Originally published 1931, *Genetics* **16:**2).
Wright, S., 1986b, The roles of mutation, inbreeding, crossbreeding and selection in evolution, in: *Evolution: Selected Papers* (W. B. Provine, ed.), Chicago University Press, Chicago. (Originally published 1932, *Proc. Sixth Intl. Congr. Genet.* **1:**356–366).
Wuethrich, B., 1998, Why sex? Putting theory to the test, *Science* **281:**1980–1982.

2

# Limits to Knowledge in Population Genetics

MICHAEL T. CLEGG

## INTRODUCTION

The modern science of biology traces, in large part, to the publication of *The Origin of Species*, by Charles Darwin (1859). By invoking natural selection as the causal agent of biological change, Darwin provided a mechanistic explanation for the elaboration of biological diversity that focused on the experimental manipulation of observable processes. In one stroke, biology was transformed from a purely descriptive enterprise concerned with cataloging the wonders of biological diversity discovered during the age of exploration to an experimental science. The second major impact of the publication was to focus late 19th-century biology on the great unsolved puzzle of hereditary transmission. It is, of course, well known that the solution to this great puzzle was published within a decade of *The Origin* but lay unappreciated for more than 30 years. It is also well known that the first two decades of the 20th century were consumed by a debate about whether the selection of minute continuous variations (as postulated by Darwin) could be consistent with the particulate system of inheritance described by Mendel.

Theoretical population genetics originated as a mathematical consistency argument to establish that natural selection could be effective under the Mendelian scheme of inheritance (Provine, 1971). This program was immensely successful because it showed that a particulate system of

---

MICHAEL T. CLEGG • Department of Botany and Plant Sciences, University of California, Riverside, California 92521.

*Evolutionary Biology, Volume 32*, edited by Michael T. Clegg *et al.*
Kluwer Academic / Plenum Publishers, New York, 2000.

inheritance was in fact essential for the conservation of genetic variance, a fundamental prerequisite for natural selection to be effective in driving evolutionary change (Fisher, 1930).

Population genetics focused on the gene as the unit of evolutionary change. Assuming particular mating schemes that determine the distribution of genotypic frequencies in each new generation and assuming various parameters describing mutation, migration, selection, and the stochastic sampling of genes transmitted to each new generation (genetic drift), equations were derived to predict the trajectory of gene frequency change. This theoretical construct continues to be elaborated to the present day. It is important to appreciate, as emphasized by Wright (1931), that the models of population genetics are at best highly simplified descriptions of the complexity of actual population dynamics. It is simply not possible to incorporate the complexities of variable selection driven by temporal and spatial environmental change together with drift, migration, and demographic structure into a comprehensive model that is susceptible to analysis. Yet, evolution is actually determined by the continuous operation of these and other factors. Thus, while the conclusions of population genetic models are mathematically true, they may not always contribute to empirical knowledge.

Experimental population genetics began in earnest in the 1930s. The initial goals were to detect selection and to estimate the intensity and pattern of selection. Other goals were to measure migration rates, to estimate effective population sizes, and to determine the overall extent of genetic variation within populations. In other words, the goals were to determine whether levels of genetic variation were sufficient for effective natural selection and to estimate the parameters that indexed the models of theoretical population genetics. The primary issue of concern was to determine the importance and strength of selection to provide empirical confirmation for theoretical calculations. Much of this program was embodied in the monumental *Genetics of Natural Populations* series initiated by Dobzhansky and his various collaborators (Dobzhansky, 1981).

This early work was largely experimental in the sense that populations were subject to manipulation and the quantities of interest were measured. Over the years it became clear that some of the parameters of interest could not be estimated, either because experimental error is too large or because parameters like effective population size and selection are hopelessly confounded (Lewontin, 1974). In more recent years, the focus has moved from manipulative experiments to historical inference. The interest in historical inference has been stimulated by molecular data that permit direct comparisons of gene sequences among species. These data are then used to infer phylogenies or to infer the evolutionary history of particular genes and gene families. There are strict limits to historical inference. It is not possible to "know" that the particular history estimated is the true history, although we may be able to say with a given probability that the true history lies within some ensemble of histories.

To summarize, mathematical population genetics is oversimplified. The theory defines the conditions for evolutionary advance, but some of the most important parameters cannot be estimated with any precision and are at best knowable in only the crudest empirical sense. Moreover, the goal of historical inference is only approximately attainable. It also is important to realize that the limits to knowledge in population genetics, and hence also the limits to knowledge in the allied sciences of molecular evolution and conservation biology, are not obvious. Workers in these fields have come to only a partial appreciation for the limits of knowledge as they have explored the empirical content of population genetic theory over a period of more than 60 years.

## A Classification of Limits to Knowledge in Population Genetics

To facilitate discussion of the limits of knowledge it is useful to consider a partial classification of the reasons why the answers to a proposition are outside the bounds of scientific knowledge. Such a classification would include the following points:

1. Knowable in principle but not with present technology.
2. Knowable in principle but the experimental error overwhelms any signal.
3. Knowable in principle but the theoretical framework for interpretation is positively misleading.
4. Unknowable because parameters are confounded.
5. Unknowable because the stochastic process that governed the history has too great a variance.
6. Unknowable because the complexity of the biological system precludes deterministic prediction.

Knowledge is often approximate and contingent on special conditions obtaining. It is rarely complete and immutable. The exploration of new knowledge occurs on a boundary that defines the currently known and unknown. This boundary is rarely sharp and well defined. In research we probe the boundary by devising new methods and new theoretical constructs that circumscribe problems and render them tractable. The objective of this chapter is to explore two contemporary issues in population genetics as a means of probing this boundary region. The first issue arises from the application of coalescence theory to the problem of inferring the historical processes that lead to a particular gene genealogy (realization). The second issue we explore is the problem of genetic–ontogenetic complexity. Specifically, I will explore the way genes interact in determining phenotype via their metabolic and developmental context.

## THE ANALYSIS OF GENE GENEALOGIES

Much of the classic theory of population genetics was framed in terms of the transmission of genes that are identical by descent. The conceptual framework allows genes to change state through mutation, but their path of descent is assumed to be known. It is then relatively easy to derive transition equations that incorporate sampling effects (genetic drift), mutation, and migration for identical-by-descent systems (see, e.g., Crow and Kimura, 1970). It is much more difficult to incorporate selection, because selection causes the systems of equations to be nonlinear and limits the scope of analysis. Until recently, the empirical technology did not permit geneticists to infer patterns of descent for genes sampled from real populations. The best that could be inferred was that two genes are identical in state, but identity in state does not imply identity by descent.

To provide a concrete example, we consider the isozyme technology of the 1960s and 1970s: A particular isozyme state is determined by the net charge and conformation of the protein molecule. (The isozyme state is defined as the location of a particular enzymatic protein on a gel medium, where the proteins have been separated by electrophoresis.) Net charge is in turn the result of both the pH of the medium and the distribution of acidic and basic amino acids comprising the protein molecule. A mutational change that alters a single amino acid may induce a location on a gel (state) that is identical to other very different amino acid configurations (Veuille and King, 1995). Hence, comigration of a band (identity in location on a gel) does not imply genetic similarity and does not permit the estimation of identity-by-descent relations. In this case, the identical-by-descent relationships are unknowable in the absence of detailed information on the breeding history of individuals in the population. The isozyme example illustrates the situation where the identity status may be knowable in principle, but it is not knowable based on the isozyme technology alone (point No. 1 above).

With today's technologies for rapid DNA sequencing, identity by descent can be estimated with considerable accuracy, linking the objects of theory more closely to the quantities that can be observed in the empirical world. This follows because DNA sequence data measure directly the mutational similarity between copies of a gene; hence, patterns of descent (genealogies) can be estimated from the joint distribution of nucleotide differences between a set of homologous sequences (sequences that have descended from a common ancestral sequence). Thus, a new technology has rendered the unknowable at least approximately knowable. An interesting further point about the technology of rapid DNA sequencing is that it yields the most elemental of all genetic data. Because these data are the most elemental (i.e., the base sequence of a gene), no further technology can be expected to provide enhanced genetic resolution. (Of course, there is much room for improving the speed and throughput rate of DNA sequencing technology, so we may expect much larger data sets in the future that will reduce uncertainty by increasing statistical power.)

Why is the identical-by-descent situation important? It is important because a great deal of information is latent in the genealogical history of a sample of alleles. A fundamental theorem of coalescence theory shows that when the genealogy is the realization of a process that is strictly a drift–mutation process, then the entire distribution of the sample history can be calculated (Hudson, 1990). More generally, however, the structure of a genealogy (estimated internode lengths, branch lengths, number of branches connecting the sample of DNA sequences) depends on the mutation rate ($\mu$), the effective population size ($N_e$), the pattern and intensity of selection, the history of recombination, and the demography of the population (Sawyer and Hartl, 1992; Hudson, 1993; Nee *et al.*, 1995; Holmes *et al.*, 1995). The distribution of the sample history cannot be calculated for this more realistic scenario. Nevertheless, comparison of the estimated sample history for a set of DNA sequences to that calculated from elementary theory provides a means of integrating over long periods of evolutionary time so that episodes of relatively weak selection can be detected (i.e., $s < 0.01$). What cannot be measured is the exact pattern of selection, recombination, and demography that actually produced the observed realization. To summarize, the analysis of gene genealogies exploits both large amounts of nucleotide sequence data and a well-defined null hypothesis of no selection ($s = 0$), but the analysis cannot provide a precise (or even very approximate) reconstruction of the actual processes that produced a given history.

Thus, the invention of a new technology appears to have brought theoretical and empirical population genetics into much closer alignment, and therefore it has moved some categories of questions from the unknowable to at least the crudely knowable. However, there are strict limits to the inferences that can be supported from these kinds of data. The limits fall into categories 2, 4, and 5 above, and before elaborating on these limits to knowledge it is useful to review in more detail the conceptual tools available for the analysis of gene genealogies.

## Inferences Based on Coalescence Theory

Many different tests of gene sequence samples have been proposed, and taken together these tests provide some basis for inferring the relative importance of the various evolutionary processes that have affected genetic variation. Under the assumption of a strict drift–mutation process (referred to below as "neutrality"), the time intervals between coalescence events (nodes of a genealogy), measured in mutational change, is dependent on $4N_e\mu = \theta$. Thus, the estimation of $\theta$ is central to testing the null hypothesis of neutrality. Assuming neutrality is accepted and assuming that the neutral mutation rate is known, the effective population size of the species can then be estimated. [The efficient estimation of $\theta$ is a nontrivial problem because the intercorrelation of sequence change has to be accounted for by estimating $\theta$ and the genealogy together. Moreover, the possibility of recombination among homologous sequences also must be built into the estimation model (Kuhner *et al.*, 1995).]

Several statistics that are simple functions of $\theta$ are used as the basis for statistical tests of a neutral genealogy. Among these statistics are $S$, the number of segregating sites, where $E(S) = a_n\ \theta$; $k$, the average number of pairwise nucleotide differences, where $E(k) = \theta$; and $\eta_s$, the number of singletons in the sample, where $E(\eta_s) = [n/(n-1)]\theta$. (Singletons are mutations that occur in only one sampled allele.) And, where $a_n = \Sigma_{i=1,n}(1/a_i)$, $n$ is the number of observations in the sample and $E$ indicates the operator for mathematical expectation. Each of these expressions is an unbiased estimator of $\theta$ (Watterson, 1975).

A test statistic introduced by Tajima (1989) is based on the difference between $k$ and $S/a_n$, and test statistics introduced by Fu and Li (1993) are based on the difference of $S/a_n$ and $\eta_s\ [(n-1)/n]$ and on the difference of $k$ and $\eta_s\ [(n-1)/n]$. Assuming neutrality, each of these test statistics has an expected value of approximately zero. However, the exact distribution of each test statistic is unknown. Critical values for statistical tests were suggested by Tajima (1989), based on assuming a beta distribution, and by Fu and Li (1993), based on simulated data. As a consequence, the actual bases for statistical analysis are very approximate.

Simonsen *et al.* (1995) have carried out extensive studies of the power of these, and several analogous test statistics, under the specific alternatives of (1) a selective sweep, (2) a population bottleneck and, and (3) population subdivision. The overall conclusion of this study was that Tajima's test is usually the most powerful test; however, the power to reject the neutral null hypothesis is limited for samples of fewer than 50 sequences of a gene, and when the sample includes 10 or fewer observations, the power is virtually nil. In the case of a selective sweep, Simonsen *et al.* (1995) note that there is a specific window of time when the selective sweep can be detected, and this corresponds to $T_s = 0.18$ to $T_s = 0.27$ (for $s = 10^{-4}$) where $T_s$ is measured in $2N_e$ generations. (Thus for $N_e = 10^6$ the range is 360,000 to 540,000 generations.) This means that many episodes of selection could occur but be undetectable because they fell outside the critical time range. In fact, for $T_s$ sufficiently large and $s = 10^{-4}$ to $10^{-2}$, the test is actually less likely to accept the alternative hypothesis of selection (even though selection is present) than the null hypothesis of no selection. The final important conclusion of the work of Simonsen *et al.* (1995) is that more statistical power is almost always obtained by sampling more individuals rather than more nucleotide sites per individual. In light of these facts, it is not surprising that the null hypothesis ($s = 0$) is rarely rejected for gene sequences sampled from within species because few studies approach sample sizes of 50 individuals. In this case, knowledge is limited by statistical power, but the issue of power can in principle be overcome with sufficiently large sample sizes (No. 2 in the above classification). Knowledge also is limited by the fact that for certain regions of the trajectory of gene frequency change, the hypotheses of selection versus neutrality are not distinguishable because the two processes are virtually identical (corresponding to No. 5 above). Finally, estimates of $N_e$ are always confounded when selection is present, because selection amounts to a restriction of effective population size (No. 4 above). There are other potential tests for selection, but all suffer from similar defects (Clegg, 1997).

## Some Empirical Data

Cummings and Clegg (1998) used the brute force approach of sampling a large number of genes to address the power limitation. This work was based on a sample of 45 sequences from the alcohol dehydrogenase 1 (*adh* 1) locus of wild barley (*Hordeum vulgare* ssp. *spontaneum*). The various test statistics described above showed significant deviations from the null hypothesis of neutrality, so we may conclude that the strictly neutral model can be rejected. However, the deviation from neutrality appeared to be accounted for by an excess of singleton amino acid replacement polymorphisms, which may indicate better conformance to the nearly-neutral hypothesis of Ohta (1973, 1992) or the background selection hypothesis of

Charlesworth *et al.* (1993), both of which posit that a flux of mildly deleterious mutations controls the distribution of polymorphism.

To interpret these data further it is important to recall that barley is a predominantly self-fertilizing plant with estimated rates of self-fertilization in the neighborhood of 99% (Brown *et al.*, 1978; Clegg *et al.*, 1978). Because self-fertilization implies high levels of homozygosis, we expect the equilibrium frequencies of deleterious mutations to be approximated by $\mu/s$, so it is surprising to find such a high frequency of replacement mutations in the sample. Estimates of effective population size from synonymous and intron sites suggests that $N_e$ is approximately $10^5$, so we would expect weak negative selection to be effective. It may be that population subdivision plays an important role in determining the distribution of replacement polymorphism. Under this assumption, selection within each small local population would be the factor determining the frequency of negatively selected variants. Some could then drift to moderate frequencies because local effective sizes may be small. Because the sample of 45 sequences was drawn from the whole geographical distribution of the species, it is not possible to test for the effects of population subdivision; however, a gross examination of the geographical distribution of variation does not reveal obvious geographical patterns.

A straightforward effort to increase statistical power has not resolved the question; instead, it has raised new questions about population substructure that require much larger sample sizes drawn from a geographically structured sample. Nor is it clear that such a sample would in fact resolve the question, because little is known of the history and demography of local populations. So the answers are elusive and regress to other subsidary questions.

What lessons may be drawn from such experiments? It is like peeling an onion. There are layers of technical innovation, each promising to relate more closely to the objects of mathematical theory. At the same time the mathematical theory is elaborated layer after layer, and yet the answers to the empirical questions that have motivated the science of population genetics remain elusive.

## ADAPTATIVE EVOLUTION IN COMPLEX BIOLOGICAL SYSTEMS

The fundamental theoretical quantity of population genetics is the gene, yet biological adaptations are not the isolated manifestation of individual genes. The target of selection is a phenotype that is determined by the interactions of many genes, both in a temporal (developmental) and in a metabolic context. The methodological triumph of classical genetics was to break out of this context dependency by studying extreme mutations that completely disrupted the function of a gene. By this means, it was possible to acquire an understanding of the "function" of a gene, but little was revealed about the subtle context-dependent changes that are thought to be the substrate of adaptive evolutionary change. In contrast, theoretical population genetics simply assumed that selection produced adaptive change and then studied the consequences of selection on single genes or occasionally on pairs of genes. Both approaches beg the central empirical question of relating adaptive phenotypic change to its underlying genetic determinants. (This corresponds to point No. 3 where knowledge is limited because the theoretical framework is misleading.)

## Chalcone Synthase Genes in Morning Glory

As a specific example of the complex interactions that arise in metabolism and development, consider the flavonoid biosynthetic pathway of plants. This biochemical pathway is responsible for flower pigmentation and it also plays a major role in plant disease defense, lignin biosynthesis, and UV protection, so the mapping from genes to phenotypes is complex. Many genes are involved in determining flower color phenotypes, but those same genes also determine other distinct aspects of phenotype that are potentially adaptive. This complexity limits our ability to dissect the adaptive contribution of any particular mutation to a single aspect of phenotype (e.g., flower color, corresponding to point No. 6 of the above classification). To make this point concrete, Fig. 1 illustrates the biochemical pathway that leads to flower pigmentation. Figure 1 also shows the various side branches that lead to other important compounds. A quick perusal of Fig. 1 is sufficient to establish that many end products (phenotypes) may be dependent on the same enzymatic steps.

To facilitate our exploration of the limits to knowledge in this system, let us take as an example ongoing analyses of the evolution of the genes encoding one enzyme in this pathway within the common morning glory, *Ipomoea purpurea.* The first key step in flavonoid biosynthesis is the condensation of three molecules of malonyl CoA with one molecule of 4-coumaroyl CoA to form the 15-carbon naringenin chalcone molecule. The condensation is catalyzed by the enzyme chalcone synthase (CHS), which typically exists as a small multigene family in plant genomes (almost all plant genes appear to exist as multigene families).

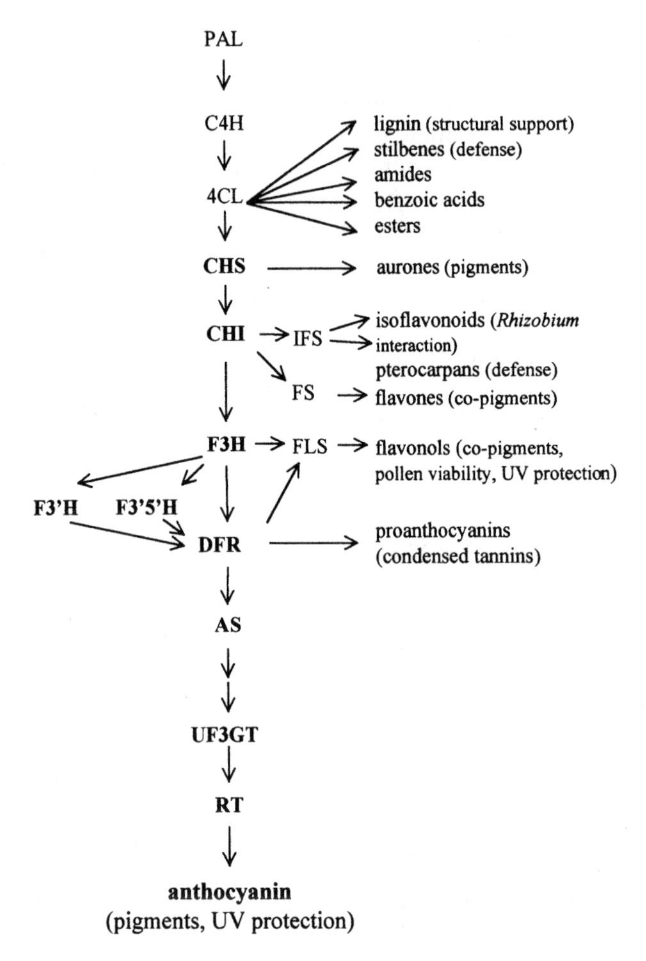


FIG. 1. A diagram of the flavonoid biosynthetic pathway.

## Gene Duplication and Divergence in Evolution

Durbin *et al.* (1995) characterized four genetic loci that encode *chs* genes in the morning glory (*chs* A, B, C, and PS). In addition, two other distantly related *chs* genes (D and E) have recently been described in the morning glory genome (Fukada-Tanaka *et al.*, 1997). The *chs* gene sequences have been determined from five *Ipomoea* species (Durbin *et al.*, 1995; unpublished observations). Four of the *Ipomoea chs* genes (A, B, C, and PS) appear to have duplicated approximately 25 million years ago, prior to the divergence of the *Ipomoea* species included in the sample. Based on genetic distances and nine characteristic amino acid substitutions, these four *chs* loci group into two related categories. Group 1 includes *chs* A and C, whereas *chs* B and PS are contained in group 2. The amino acid substitutions observed in group 2 are characteristic of stilbene synthases (STS) and may indicate a functional divergence between duplicate *chs* genes. The independent evolution of the STS function from CHS has occurred several times in flowering plant evolution (Tropf *et al.*, 1994), and *chs* genes have evidently acquired other functions in flowering plant evolution as well (Durbin *et al.*, 2000). Genealogical relationships among duplicate *chs* genes are illustrated in Fig. 2, where presumptive shifts in enzymatic function are highlighted.

The *Ipomoea chs* genes appear to be evolving at a rapid rate and have a relatively low ratio of amino acid replacement to synonymous substitutions (about 5.6 synonymous to replacement changes) (Glover *et al.*, 1996; Durbin *et al.*, 2000). Synonymous substitutions are presumed to be selectively neutral; however, replacement substitutions are more directly exposed to selection by virtue of their associated amino acid changes. Thus, the ratio of synonymous to replacement substitution serves as an index of the level of selective constraint operating on sequences, with a lower ratio implying either reduced selective constraint or diversifying selection. At this stage we can only speculate on the potential adaptive bases for *chs* gene family expansion, but divergence in enzymatic function and divergence in developmental expression patterns appear to be common characteristics of *chs* gene family evolution.

The expression of the *chs* gene family members in *I. purpurea* is differentially regulated (Durbin *et al.*, 2000). Four of the genes (*chs* A, B, C, and D) are known to be expressed based on the recovery of cDNA clones corresponding to each gene. However, *chs* A is found in the limb of the flower corolla, whereas *chs C* is expressed in the throat of the corolla. The *chs* mRNA also is found in mature anther tissue, although it is not established whether expression is specific to either pollen or anther tissue. Finally, *chs* expression occurs in the stem, pistil, cotyledon, and sepal tissues but has not been observed in roots of young seedlings. These data are consistent with the multiple biological functions of the products of flavonoid biosynthesis. They also reveal considerable evolutionary divergence in developmental expression during the time since the elaboration of this gene family. We assume that diverging patterns of developmental expression are the result of adaptive responses to environmental conditions.

To summarize, the *chs* genes in *Ipomoea* have duplicated and diverged in function, and they have diverged in specific expression patterns during *Ipomoea* evolution. All the evidence indicates that some of the *chs* genes have been subject to adaptive evolution, but the specific phenotypic effects

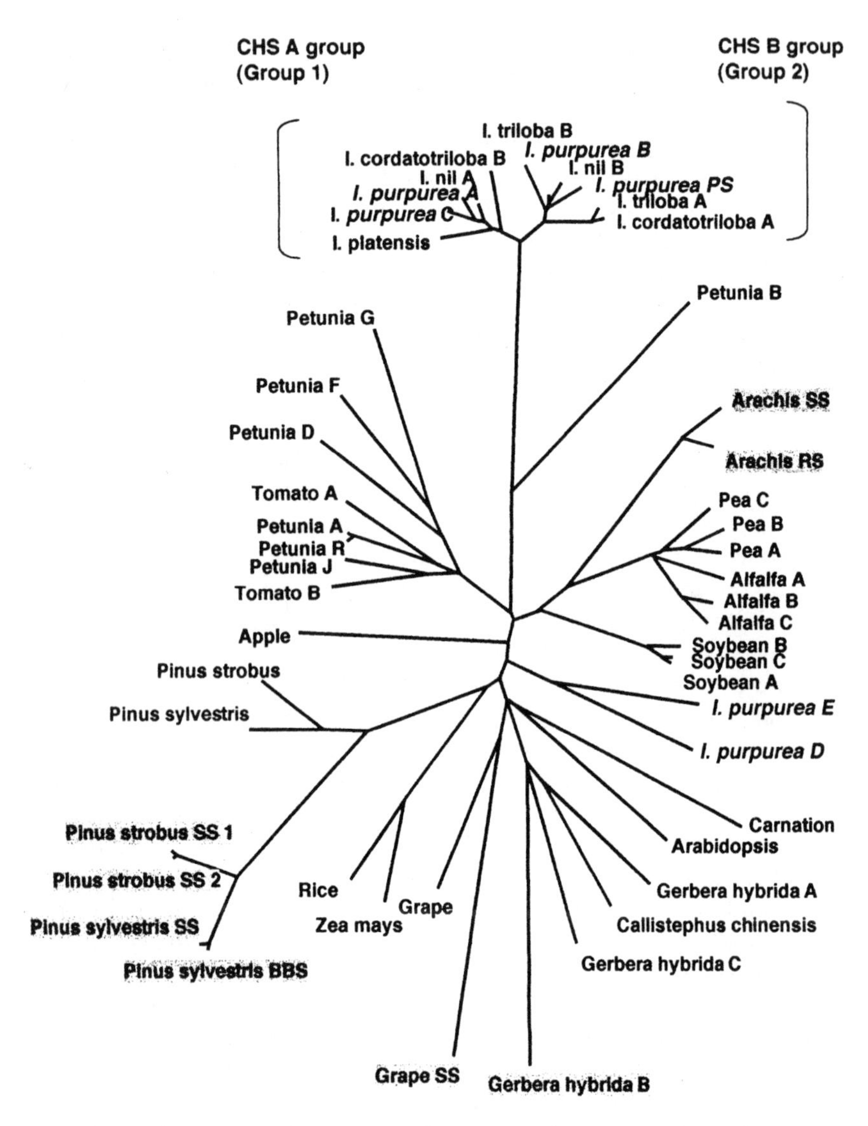


FIG. 2. A neighbor-joining tree of chalcone synthase gene relationships.

associated with *chs* evolution remain obscure. We may have reached the limits of knowledge regarding the relationship between genetic change and adaptive phenotypic change in this system, but maybe not. It may be necessary to probe in more detail associated changes in the other genes of flavonoid biosynthesis that are functionally linked to *chs* in determining phenotype.

## Interlocus Gene Conversion–Recombination

Huttley *et al.* (1997) sampled and sequenced *chs* A genes from 14 lines that originated from collections in Mexico. (The center of origin of *I. purpurea* is in the highlands of central Mexico.) Four lines from Georgia and North Carolina, where the plant is an introduced weedy species, were also sampled. No nucleotide sequence diversity was detected at *chs* A in the very limited sample from the United States. The Mexican-derived materials had much higher levels of nucleotide sequence diversity, with 11 distinct haplotypes. Further examination of the *chs* A sequences reveals that the majority of variation resides in exons, as indicated by higher $\theta$ estimates for exon comparisons minus the flanking sequences and introns. Finally, a comparison between the *chs* A allele polymorphic sites indicates that at many of these sites one of the nucleotide states is present in non-*chs* A gene family members, and these observations strongly suggest that both the high level of nucleotide diversity present in the *I. purpurea chs* A genealogy and the relatively low ratio of synonymous to nonsynonymous substitutions between *chs* A alleles are probably derived from low to moderate rates of interlocus recombination–gene conversion among the different *chs* gene family members (Huttley *et al.*, 1997). This kind of observation is not unprecidented (see Chapter 8, this volume), but it does represent a source of variation that is outside classical population genetic theory. In this sense our empirical knowledge has been expanded through the discovery of phenenoma that were not anticipated by population genetics theory. But the question remains: Can we determine the precise causal pathways through which adaptive evolutionary changes are mediated or is this beyond the methodological reach of contemporary biology?

## Genetics and Population Biology of White Flower Color Phenotypes in Morning Glory

Genes that modify the mating system by increasing rates of self-fertilization have been objects of theoretical interest because they are selected by virtue of their control of the system of genetic transmission (Fisher, 1941; Holsinger, 1996). Extensive experiments with the common morning glory over a period of many years have shown that insect pollinators discriminate against white flowers, which as a consequence show higher rates of self-fertilization relative to fully pigmented flowers (Clegg and Epperson, 1988). A deeper analysis of selection on white flower phenotypes is complex and requires consideration of pollen discounting, maternal seed fertility, inbreeding depression among seed derived from self-fertilization relative to outcrossing, and the effect of flower color frequency on the likelihood of self-fertilization (Holsinger, 1996). So far, a completely comprehensive analysis has not been provided because the statistical power to detect modest selection ($s < 0.05$) is limited when the scale of experiments is of a practical size.

The genetic situation is also complex because more than one locus may lead to a white flower color phenotype. For instance, a mutant phenotype that yields an albino flower (A/a locus) (Epperson and Clegg, 1988) is characterized by the absence of the *chs* D transcript, which suggests that *chs* D is uniquely responsible for the absence of anthocyanin pigments in the aa phenotype responsible for flower color (Durbin *et al.*, 2000). However, the albino aa genotype is rare in populations (Epperson and Clegg, 1992), and a second locus (W/w) (Epperson and Clegg, 1988) that yields a white flower color phenotype with pigmented floral rays is also discriminated against by pollinators. The W/w locus appears to be a regulatory locus based on genetic analysis (Ennos and Clegg, 1983) and on expression studies (unpublished observations).

Selection on flower pigmentation by pollinators is about as simple a situation to analyze as possible, because the genetics and molecular biology are susceptible to analysis and the environmental agent of selection (bumblebees in this case) can be isolated and studied. Even in this case, however, the analysis is complicated by the fact that more than one gene is responsible for the phenotypes discriminated against in nature by the insect pollinators that affect plant reproduction. It also is complicated by considerations of experimental power. It is very probable that deeper analyses will reveal additional complexities in the causal pathway between selection and phenotypic determination.

It is probable that the limitations to present knowledge are simply the consequence of inadequate experimental data, and it seems reasonable to expect that further analyses will inform and enrich our understanding of the causal pathways linking phenotypic diversity to adaptation. Nevertheless, the study of adaptation involves analyses at multiple levels of biological organization where pathways of causal determination are imbedded in networks of gene interaction and ecological interaction. It is clear that these networks are very difficult to penetrate and to unravel in a causal sense.

## SUMMARY

Initial conditions can play a determining role in the evolution of systems of knowledge just as they play a determining role in other systems (e.g., Chapter 1, this volume). In the case of population genetics the initial conditions focused on the analysis of the transmission of abstract entities (genes) subject to highly simplified dynamical systems. This strategy provided the mathematical argument that was required for the reconciliation of Darwinian selection and Mendelian heredity, but it also guided and determined subsequent empirical research. This rich theoretical structure has been essential to the development of a rigorous hypothesis-testing framework, and it has been especially valuable in areas where historical inference is the paramount objective.

Theory has been less successful when we move from the abstract to the concrete. Thus, for example, an empirical understanding of selection is much richer than implied in the simple summary parameters of population genetic theory. This is especially apparent in the study of adaptation, where the essential problem is to unravel a complex system and by so doing to understand its internal workings. In contrast, the strategy of model building is to simplify by ignoring internal complexity for the sake of generality. Both approaches are valid and important, but both also circumscribe knowledge and impose real limits. One thesis of this chapter is that important limits to knowledge lie in the region of disjunction between theory and empirical research in population genetics.

I argued above that the study of adaptation involves a translation between different levels of biological organization: the ecological, the developmental, and the genetic levels. Moreover, the study of adaptation involves spatial interactions (e.g., pollinator flight movements in the transmission flower color genes) and long temporal pathways that cannot be observed in human time scales (e.g., the elaboration of gene family diversity and divergence with respect to chalcone synthase genes). Given this complexity, it seems appropriate to ask whether it is possible to unravel the precise causal pathways through which adaptive change occurs. The reality of biological systems is not linear; instead, evolutionary change is determined by networks of interaction where context may be everything. These networks may be loosely connected at the boundaries that characterize different levels of biological organization. The translation across these boundaries is very poorly understood and seldom studied in contemporary biological science. We do not know how knowledge may be limited by the multiple layers of organization that exist in and are confined by a spatial and temporal context. It does seem evident, however, that the traditional methods of description are inadequate to this larger problem. Certainly future biological research will be increasingly concentrated on these boundary issues. It may be that the paradigms governing biological research will need to be restructured in response to this imperative (as suggested in Chapters 4 and 5, this volume). It is clearly evident that there are limits to knowledge imposed by the structure of current research paradigms. An exploration of these limits is an essential prerequisite to the further elaboration of evolutionary science.

## Acknowledgments

I thank Dr. Michael Cummings for comments on the manuscript and for stimulating discussion about this project. I also thank Professor Ward Watt for a helpful exchange of ideas. This work was supported in part by the Alfred P. Sloan Foundation.

## REFERENCES

Brown, A. H. D., Zohary, D., and Nevo, E., 1978, Outcrossing rates and heterozygosity in natural populations of *Hordeum spontaneum* Koch in Israel, *Heredity* **41:**49–62.
Charlesworth, B., Morgan, M. T., and Charlesworth, D., 1993, The effect of deleterious mutations on neutral molecular evolution, *Genetics* **134:**1289–1303.
Clegg, M. T., 1997, Plant genetic diversity and the struggle to measure selection, *J. Hered.* **88:**1–7.
Clegg, M. T., and Epperson, B. K., 1988, Natural selection on flower color polymorphisms in morning glory populations, in: *Plant Evolutionary Biology* (L. Gottlieb and S. K. Jain, eds.), pp. 255–273, Chapman-Hall Ltd., London.
Clegg, M. T., Kahler, A. L., and Allard, R. W., 1978, The estimation of life cycle components of selection in an experimental population of barley, *Genetics* **89:**765–792.
Crow, J. F., and Kimura, M., 1970, *An Introduction to Population Genetics Theory*, Harper & Row, New York.
Cummings, M. P., and Clegg, M. T., 1998, Nucleotide sequence diversity at the alcohol dehydrogenase I locus in wild barley (*Hordeum vulgare* ssp. *spontaneum*): An evaluation of the background selection hypothesis, *Proc. Natl. Acad. Sci. USA* **95:**5637–5642.
Darwin, C., 1859, *The Origin of Species*, John Murray, London.
Dobzhansky, T., 1981, *Dobzhansky's Genetics of Natural Populations I-XLIII* (R. C. Lewontin, J. A. Moore, W. B. Provine, and B. Wallace, eds.), Columbia University Press, New York.
Durbin, M. L., Learn, G. J., Huttley, G. A., and Clegg, M. T., 1995, Evolution of the chalcone synthase gene family in the genus *Ipomoea, Proc. Natl. Acad. Sci. USA* **92:**3338–3342.
Durbin, M. L., McCaig, B., and Clegg, M. T., 2000, Molecular evolution of the chalcone synthase multigene family in the morning glory genome, *Plant Mol. Biol.* **42:**79–92.
Ennos, R. A., and Clegg, M. T., 1983, Flower color variation in morning glory, *Ipomoea purpurea* Roth, (Convolvulaceae), *J. Hered.* **74:**247–250.
Epperson, B. K., and Clegg, M. T., 1988, Genetics of flower color polymorphism in the common morning glory, *Ipomoea purpurea, J. Hered.* **79:**64–68.
Epperson, B. K., and Clegg, M. T., 1992, Unstable white flower color genes and their derivatives in the morning glory, *J. Hered.* **83:**405–409.

Fisher, R. A., 1930, *The Genetical Theory of Natural Selection*, Oxford University Press, Oxford, England.

Fisher, R. A., 1941, Average excess and average effect of a gene substitution, *Ann. Eugen.* **11:**53–63.

Fu, Y.-X., and Li, W.-H., 1993, Statistical tests of neutrality of mutations, *Genetics* **133:**693–709.

Fukada-Tanaka, S., Hoshino, A., Hisatomi, Y., Habu, Y., Hasebe, M., and Lida, S., 1997, Identification of new chalcone synthase genes for flower pigmentation in the Japanese and common morning glories, *Plant Cell Physiol.* **38:**754–758.

Glover, D., Durbin, M. L., Huttley, G., and Clegg, M. T., 1996, Genetic diversity in the common morning glory, *Plant Species Biol.* **11:**41–50.

Holmes, E. C., Nee, S., Rambaut, A., Garnett, G. P., and Harvey, P. H., 1995, Revealing the history of infectious disease epidemics using phylogenetic trees, *Phil. Trans. R. Soc. Lond. B* **349:**33–40.

Holsinger, K. E., 1996, Pollination biology and the evolution of mating systems in flowering plants, *Evol. Biol.* **29:**107–149.

Hudson, R., 1990, Gene genealogies and the coalescent process, *Oxford Surv. Evol. Biol.* **7:**1–44.

Huttley, G. A., Durbin, M. L., Glover, D. E., and Clegg, M. T., 1997, Nucleotide polymorphism in the chalcone synthase-A locus and evolution of the chalcone synthase multigene family of common morning glory (*Ipomoea purpurea*), *Mol. Ecol.* **6:**549–558.

Kuhner, M. K., Yamato, J., and Felsenstein, J., 1995, Estimating effective population size and mutation rate from sequence data using Metropolis-Hastings sampling, *Genetics* **140:**1421–1430.

Lewontin, R. C., 1974, *The Genetic Basis of Evolutionary Change*, Columbia University Press, New York.

Nee, S., Holmest, E. C., Rambau, A., and Harvey, P. H., 1995, Inferring population history from molecular phylogenies, *Phil. Trans. R. Soc. Lond. B* **349:**25–31.

Ohta, T., 1973, Slightly deleterious mutant substitutions in evolution, *Nature* **246:**96–98.

Ohta, T., 1992, The nearly neutral theory of molecular evolution, *Annu. Rev. Ecol. Syst.* **23:**263–286.

Provine, W. B., 1971, *The Origins of Theoretical Population Genetics*, University of Chicago Press, Chicago.

Sawyer, S. A., and Hartl, D. L., 1992, Population genetics of polymorphism and divergence, *Genetics* **132:**1161–1176.

Simonsen, K. L., Churchil, G. A., and Aquadro, C. F., 1995, Properties of statistical tests of neutrality for DNA polymorphism data, *Genetics* **141:**413–429.

Tajima, F., 1989, Statistical method for testing the neutral mutation hypothesis by DNA polymorphism, *Genetics* **123:**585–595.

Tropf, S., Lanz, T., Rensing, S. A., Schroder, J., and Schroder, G., 1994, Evidence that stilbene synthases have developed from chalcone synthases several times in the course of evolution, *J. Mol. Evol.* **38:**610–618.

Veuille, M., and King, L. M., 1995, Molecular basis of polymorphism at the esterase-5B locus in *Drosophila pseudoobscura, Genetics* **141:**255–262.

Watterson, G. A., 1975, On the number of segregating nucleotide sites in genetical models without recombination, *Theor. Pop. Biol.* **7:**256–276.

Wright, S., 1931, Evolution in Mendelian populations, *Genetics* **16:**97–159.

# II

# The Philosophy of Biology, Paradigms, and Paradigm Shifts

The philosopher of science, Alex Rosenberg (Chapter 3), begins this section with a provocative chapter that makes two major claims. The first is that prediction in biology is severely limited, and the second is that the only law in biology is the law of natural selection. These two claims are related. Rosenberg argues that selection is blind to structure and selects only on function, because any functional solution to a particular adaptive challenge is sufficient. It is thus not possible to predict the structural solutions to biological problems of future adaptation. What of the mathematical models of population genetics that appear to capture general deductive truths about genetic transmission? Rosenberg argues that these deductive systems are largely unconnected to empirical reality because they are built from a highly simplified version of biological processes. So we cannot expect accurate prediction from models that are mere parodies of reality.

Rosenberg goes on to consider the role of historical inference in evolutionary biology. In this context, I need to add a word of explanation for one passage where he cites a statement of mine that begins with "our strategy in exploring limits to knowledge is to study the particular . . ." This passage did not survive into the final version of the chapter that I authored for the volume. The passage was included in an earlier version circulated to the conference participants, and it reflects my strategy for proposed research. However, the passage seemed inappropriate for the final chapter and was removed, but Rosenberg's critique of this passage is relevant and is included here. In this critique, Rosenberg takes on the larger question of the value of historical inference in biology. He asks, "Why do we want to know history?" He answers this question by noting that many biological theories can only be tested retrospectively, and further that the law of natural selection, while transtemporarily exceptionless, produces outcomes that are dependent on initial conditions. Hence, our current position can only be understood through reference to some initial starting point.

The biologist Ward Watt (Chapter 4) rejects the notion of ultimate limits to knowledge in evolutionary biology. He argues that the modern synthesis produced an "amechanistic" view in evolutionary biology, in which the study of biological mechanisms is subordinated to the statistical view of genetic change encapsulated in the theory of population genetics. Watt observes that the amechanistic paradigm represented by the modern synthesis emphasizes simple additive relationships between phenotype and genotype and deemphasizes the study of biological interaction. In contrast, careful experimental research with well-defined but complex phenotypes, such as bristle number in *Drosophila*, reveal instead a rich field of interaction among the relatively small number of genes that underlie the bristle number phenotype. On the basis of these examples, Watt calls for a new experimental paradigm of phenotypic analysis built on the central tenet of adaptation, and he provides case studies to bolster his argument.

Jody Hey (Chapter 5) also rejects the notion of limits to knowledge in evolutionary biology. Hey notes that imprecision of language can be the basis for limits to understanding in many areas, and he argues that our linguistic precision improves as a problem is elaborated. He goes on to consider scientific revolutions and uses the development of the theory of electromagnetism by James Clerk Maxwell as an example. Hey argues that even though Maxwell's work revolutionized the physics of his time, Maxwell was unable to predict and appreciate the many ways his theories would unfold in the future. On the basis of this analogy, Hey cautions us to be wary of accepting our current knowledge as secure. He then speculates that a revolution in evolutionary genetics may be in progress, owing to the frequent statistical rejection of neutrality when population samples of DNA sequence data are analyzed. One especially intriguing rejection arises in the study of codon bias, where very weak selection intensities must be effective. Hey argues that contemporary data are consistent with weak selection affecting most segregating nucleotide sites, owing to the confounding of linkage, selection, and effective population size. The end result is a reduction in the efficacy of natural selection. Hey concludes that the selection–drift dichotomy of classic population genetics theory is misleading, and he suggests that a mature theory will emerge to reconcile this dichotomy.

It is interesting to contrast the views of the biologists Watt and Hey. Both see evolutionary genetics as being restricted by theoretical constructs that do not engage the interesting problems presented by contemporary data. Both are optimistic and believe that new theory and a broader experimental paradigm will be elaborated to accommodate the growth of evolutionary genetics. Neither see current or future limits to knowledge in evolutionary biology. In contrast, the philosopher Rosenberg sees the structure of the science as limiting. Why the difference? Biologists are concerned with the particular. How did a particular adaptation arise, what is its genetic basis, what features of the environment selected for this adaptation? The philosopher is interested in the general. What is the structure of the science and how does its structure limit explanatory power?

# 3

# Laws, History, and the Nature of Biological Understanding

ALEX ROSENBERG

Michael Clegg, in the Preface to this volume, notes the role biology is expected to play in application to policy:

> I was stimulated to consider the limits to knowledge in the applied science of conservation biology as Chair of the National Research Council Committee on Science and the Endangered Species Act. The theoretical foundations of ecology and population genetics play a major role in the science of biological conservation. For instance, theory, in conjunction with empirical information, informs approaches to reserve design and to the elaboration of breeding strategies. The fundamental science of evolutionary biology is called on by managers in deciding when a species is no longer in danger of extinction and when a listing is the appropriate action. (p. viii)

Applying theory to make decisions about the future is prediction. In essence, Clegg has focused our attention on the obstacles to prediction in the biological sciences. Each of the differing kinds of limits to biological knowledge that he catalogs are, in effect, limits on our ability to predict how data and observations will turn out. The application of biology to practical matters of policy is but one arena in which these limits to biological knowledge are manifest. But they are equally evident in the development of biological theory, for in this arena too, development, improvement, or correction is a matter of predicting observations and then revising them to accommodate divergence of data from theory.

---

ALEX ROSENBERG • Department of Philosophy, Duke University, Durham, North Carolina 27706.

*Evolutionary Biology, Volume 32*, edited by Michael T. Clegg *et al.*
Kluwer Academic / Plenum Publishers, New York, 2000.

Predictions do not have to be about the future of course. They can be made for the as-yet-untraced past. We may then seek data that record the past and that test these predictions—or rather retrodictions. But unless our interest is purely historical, these retrodictions are but means to other ends. And these other ends or goals that scientific theorizing serves are twofold: one is practical application to future policy, as Clegg's conclusion reflects; but the second is a fundamental understanding of the way the world works. This second goal is the search for general theoretical understanding reflected in the discovery of the fundamental laws of working that govern biological processes. Without such laws, biological theory will turn out to be limited in the most severe of the six ways Clegg notes. Answers to biology's questions will be unknowable because for reasons of complexity or indeterminism the biological system precludes prediction (see Chapter 2, this volume). If this is the case, then biology will turn out to be history, and not simply because history provides the data for retrodiction, but because biological understanding will turn out to be limited, like human history, to tracing out causal pathways in the past but not the future.

In this chapter I argue that biology is in fact so limited. My argument is not a philosophical one. It is an argument from biology. That is, I argue from the nature of selection to the conclusion that there are no laws to be found in biology beyond those of the theory of natural selection. That is, there are no laws in systematics, development, physiology, or for that matter genetics—population or molecular. Or at least there are no biological laws that could be uncovered by agents of our cognitive and computational capacities. And it is in the nature of the mechanism of evolution that there can be no such laws. This, I would argue, is the impenetrable limit to our biological knowledge. Once I have explained how this limit operates, I will raise some fundamental and as yet unanswered questions about the nature of biological laws that this limitation raises.

Natural selection "chooses" variants by some of their effects, those that fortuitously enhance survival and reproduction. When natural selection encourages variants to become packaged together into larger units, the adaptations become functions. Selection for adaptation and function kicks in at a relatively low level in the organization of matter. As soon as molecules develop the disposition chemically, thermodynamically, or catalytically to encourage the production of more tokens of their own kind, selection gets enough of a toehold to take hold. Among such duplicating molecules, at apparently every level above the polynucleotide, there are frequently to be found multiple physically distinct structures with some (nearly) identical rates of duplication, different combinations of different types of atoms and molecules, that are close enough to being equally likely to foster the appearance of more tokens of the types they instantiate. Thus, so far as adaptation is concerned, from the lowest level of organization onward there are frequently ties between structurally different molecules for first place in the race to be selected. And, as with many contests, in case of ties, duplicate prizes are awarded. For the prizes are increased representation of the selected types in the next "reproductive generation." And this will be true up the chain of chemical being all the way to the organelle, cell, organ, organism, kin group, and so on.

It is the nature of any mechanism that selects for effects that it cannot discriminate between differing structures with identical effects. And functional equivalence combined with structural difference will always increase as physical combinations become larger and more physically differentiated from one another. Moreover, perfect functional equivalence is not necessary. Mere functional similarity will do. So long as two or more physically different structures have packages of effects, each of which has roughly the same repercussions for duplication in the same environment, selection will not be able to discriminate between them unless rates of duplication are low and the environment remains very constant for long periods of time. Note that natural selection makes functional equivalence-*cum*-structural diversity the rule and not the exception at every level of organization above the molecular. In purely physical or chemical processes that produce very different molecules as outputs for a given molecule as input, there is no opportunity for nature to select by effects; structural differences with equivalent effects are the exception, if they obtain at all.

Now, if selection for function is blind to differences in structure, then there will be nothing even close to a strict law in any science that individuates kinds by selected effects, that is by functions. This will include biology and all the special sciences that humans have elaborated. From molecular biology through neuroscience, psychology, sociology, economics, and so forth, individuation is functional. That cognitive agents of our perceptual powers individuate functionally should be no surprise. Cognitive agents seek laws relating natural kinds. Observations by those with perceptual apparatus like ours reveal few immediately obvious regularities. If explanations require generalizations, we have to theorize. We need labels for the objects of our theorizing even when they cannot be detected because they are too big (galaxy), or too small (quark), or mental (pain). We cannot individuate electrons, genes, ids, expectations about inflation, or social classes structurally because we cannot detect their physical features. But we can identify their presumptive effects. This makes most theoretical vocabulary "causal role" descriptions.

It is easy to show that there will be no strict exceptionless generalizations incorporating functional kinds. Suppose we seek a generalization about all *F*s, where *F* is a functional term, like gene, or wing, or belief, or clock, or prison, or money, or subsistence farming. We seek a generalization of the form

$$(x)[Fx \to Gx]$$

In effect our search for a law about *F*s requires us to frame another predicate, *Gx*, and determine whether it is true of all items in the extension of *Fx*. This new predicate, *Gx*, will itself either be a structural predicate or a functional one. Either it will pick out *G*s by making mention of some physical attribute common to them or *Gx* will pick out *G*s by descriptions of one or another of the effects (or just possibly the causes) that everything in the extension of *Gx* has. Now, there is no point seeking a structural, physical feature all members in the extension of *Fx* bear; the class of *Fx*s is physically heterogeneous just because they all have been selected for their effects. It is true that we may find some structural feature shared by most or even all of the members of *F*. But it will be a property shared with many other things—like mass or electrical resistance—properties that have little or no explanatory role with respect to the behavior of members of the extension of *Fx*. For example, the exceptionless generalization that "all mammals weigh less than $10^{2,000,000,000}$ grams" does relate a structural property—weight—to a functional one—mammality—but this is not an interesting biological law.

Could there be a distinct functional property different from *F* shared by all items in the extension of the functional predicate *Fx*? The answer must be highly improbable. If *Fx* is a functional kind, then the members of the extension of *Fx* are physically diverse, owing to the blindness of selection to structure. Since they are physically different, any two *F*s have nonoverlapping sets of effects. If there is no item common to all these nonoverlapping sets of effects, selection for effects has nothing to work with. It cannot uniformly select all members of *F* for some further adaptation. Without such a common adaptation, there is no further function all *F*s share in common.

Whether functional or structural, there will be no predicate *Gx* that is linked in a strict law to *Fx*. We may conclude that any science in which kinds are individuated by causal role will have few if any exceptionless laws. So long as biology and the special sciences continue so to individuate their kinds, the absence of strict laws will constitute a limitation on science. How serious a limitation will this be?

Not much, some philosophers of science will say. These philosophers hold that many *ceteris paribus* generalizations do have explanatory power, and that they do so because they bear nomological force, for all their exceptions and exclusions. A defense of the nomological status and explanatory power of exception-ridden generalization may even be extended to the claim that many generalizations of the physical sciences are themselves bedecked with *ceteris paribus* clauses; accordingly, the ubiquity of such generalizations in biology is no special limitation on its scientific adequacy.

Whether or not there are nonstrict laws in physics and chemistry, there is a good argument for thinking that the exception-riven generalizations—the "nonstrict laws" of biology—will not be laws at all. For the existence of nonstrict laws in a discipline requires strict ones in that discipline or elsewhere to underwrite them. If there are no strict laws in a discipline, there can be no nonstrict ones in it either.

The trouble with inexact *ceteris paribus* laws is that it is too easy to acquire them. Because their *ceteris paribus* clauses excuse disconfirming instances, we cannot easily discriminate *ceteris paribus* laws with nomological force from statements without empirical content maintained come what may and without explanatory force or predictive power. Now the difference between legitimate *ceteris paribus* laws and illegitimate ones must turn on how they are excused from disconfirmation by their exceptions. Legitimate *ceteris paribus* laws are distinguished from vacuous generalizations because the former are protected from disconfirmation through apparent exceptions by the existence of independent interfering factors. An interfering factor is independent when it explains phenomena distinct from those the *ceteris paribus* law is invoked to explain. This notion of an independent interferer can be explicated in the following terms:

A law of the form,

$$(1)\ \textit{ceteris paribus},\ (x)(Fx \rightarrow Gx)$$

is nonvacuous, if three conditions are filled: $Fx$ and $Gx$ do not name particular times and places, since no general law can do this; there is an interfering factor, $Hx$, which is distinct from $Fx$, and when $Gx$ does not co-occur with $Fx$, $Hx$ explains why; and $Hx$ also explains other occurrences that do not involve the original law $(x)(Fx \rightarrow Gx)$.

More formally, and adding some needed qualifications:

1. The predicates $Fx$ and $Gx$ are nomologically permissible.
2. $(x)(Fx \rightarrow Gx$ or $(EH)$ [$H$ is distinct from $F$ and independent of $F$, and $H$ explains not-$Gx$) or $H$ together with $(x)(Fx \rightarrow Gx)$ explains not-$Gx$].

3. $(x)(Fx \rightarrow Gx)$ does sometimes explain actual occurrences (i.e., the interfering factors, $H$, are not always present), and $H$ sometimes explains actual occurrences [i.e., $H$ is not invoked only when there are apparent exceptions to (1)].

So, *ceteris paribus* laws are implicitly more general than they appear. Every one of them includes a commitment to the existence of independent interferers.

This conception of *ceteris paribus* laws captures intuitions about such laws that have been expressed repeatedly in the philosophy of science. But note that on this account of nonstrict laws there will have to be some laws in some discipline that are strict after all. If all the laws in a discipline are *ceteris paribus* laws, then the most fundamental laws in that discipline will be *ceteris paribus* laws as well. But then there are no more fundamental laws in the discipline to explain away its independent interferers. Unless there are strict laws somewhere or other to explain the interferers in the nonstrict laws of the discipline, its nonstrict statements will not after all qualify as laws at all. Without such explainers, the nonstrict laws will be vulnerable to the charge of illegitimacy if not irretrievable vagueness, explanatory weakness, and predictive imprecision. Or they will be singular statements describing finite sets of casual sequences. They will fail the test for being a nonvacuous *ceteris paribus* law.

Of course, if the fundamental laws in a discipline are not *ceteris paribus* but are strict laws, this problem will not arise. This, I would argue, is indeed the case in physics and atomic chemistry for that matter. The Schrodinger wave equation does not obtain *ceteris paribus*, nor does light travel at constant velocity *ceteris paribus*. But as we have seen, in biology functional individuation precludes strict laws; there are no nonstrict ones either. Unless of course physical science can provide the strict laws to explain away the independent interferers of the nonstrict laws of biology. Alas, as we shall see, this is not an outcome that is in the cards. But generalizations that are neither strict nor nonstrict laws do not have the nomological force that scientific explanation requires. If biological explanations in which these generalizations figure do explain, they do so on a different basis from that of the rest of natural science.

Of course, biology could circumvent limits on the discovery of laws governing the processes it treats, if it were to forego functional individuation. By identifying the kinds about which we theorize and predict structurally, it could avoid the multiple realization problem bequeathed by the conjunction of functional individuation and natural selection.

But this conclusion hardly constitutes methodological advice anyone is likely to follow in biology or elsewhere beyond physical science. The reason is that foregoing functional individuation is too high a price to pay for laws: the laws about structural kinds that creatures like us might uncover will be of little use in real time prediction or intelligible explanation of phenomena under descriptions of interest to us.

What would it mean to give up functional individuation? If you think about it, most nouns in ordinary language are functional; in part this preponderance is revealed by the fact that you can verb almost any noun these days. And for reasons already canvassed, most terms that refer to unobservables are functional as well, or at least pick out their referents by their observable effects. What is more to the point, the preponderance of functional vocabulary reflects a very heavy dose of anthropomorphism, or at least human interests. It is not just effects, or even selected effects that our vocabulary reflects, but selected effects important to us either because we can detect them unaided and/or because we can make use of them to aid our survival. We cannot forego functional language and still do much biology about phenomena above the level of the gene and protein. "Plants," "animal," "heart," "valve," "cell," "membrane," "vacuole"—these are all functional notions. Indeed, "gene" is a functional notion. To surrender functional individuation is to surrender biology altogether in favor of organic chemistry.

It is a fact about the history of biology that contributions to predictive success particularly the prediction of new phenomena as opposed merely to prediction of particular events, are almost entirely unknown outside of the most fundamental compartments of molecular biology, compartments where functional individuation is limited and nature has less scope to accomplish the same end by two or more different means. The startling predictions of new phenomena that have secured theoretical credibility, for example, the prediction that DNA bases code for proteins, emerge from a structural theory par excellence: Watson and Crick's account of the chemical structure of DNA. Even where relatively well-developed theory has inspired prediction of hitherto undetected phenomena, these predictions, as noted above, have characteristically been disconfirmed beyond a limited range. Beyond molecular biology, the explanatory achievements of biological theory are not accompanied by a track record of contributions to predictive success, with one exception.

The only generalizations in biology that have consistently led to the discovery of new phenomena are those embodied in the theory of natural selection itself. Though sometimes stigmatized as Panglossian in its commitment to adaptation, evolutionary theory's insistence on the universal relentlessness of selection in shaping effects into adaptations repeatedly has led to the discovery of remarkable and unexpected phenomena. This should be no surprise, since the theory of natural selection embodies the only set of laws—strict or nonstrict—to be discovered in biology. It is because they are laws that they figure in my empirical explanation of why there are no (other) laws in the discipline.

By and large, biologists and philosophers have accepted that generalizations connecting functional kinds are not laws of the sort we are familiar with from physical science; but instead, going on to rethink the nature of biology, this conclusion has led them into philosophy. They have sought to redefine the concept of scientific law to accommodate the sort of non-nomological generalizations that biological explanations do in fact appeal to. But doing so incurs an obligation to provide an entirely new account of the nature of biological understanding—to see why we need to consider the nature of theoretical models in biology in the absence of general laws.

As with other sciences, in the absence of general theory, biologists construct and apply models to explain and organize data. In the physical sciences, these models are expected eventually to give way to general theory; that is, sets of laws that work together to explain why the models work well in some cases and poorly in others. The models that characterize evolutionary biology do not work like this, as Clegg notes in Chapter 2 (Section 1) in this volume:

> It is important to appreciate, as emphasized by Wright (1931), that the models of population genetics are at best highly simplified descriptions of the complexity of actual population dynamics. It is simply not possible to incorporate the complexities of variable selection driven by temporal and spatial environmental change together with drift, migration, and demographic structure into a comprehensive model that is susceptible to analysis. . . . Thus, while the conclusions of population genetic models are mathematically true, they may not always contribute to empirical knowledge. (p. 36)

This last observation of Clegg's is crucial. Biological models are not empirical generalizations, not even weak ones. One classic example of the character of biological models is R. A. Fisher's (1957) model for the evolutionary stability of the sex ratio. Fisher's elegant model applies everywhere and always, if there are two sexes, selection, and no interference. Like a universal law, Fisher's model is a mathematical truth without empirical content.

Of course every science embodies purely mathematical truths; physics exploits Euclidean and non-Euclidean models of space. But there is more to physics than mathematical truths. Its models—mathematical and otherwise—are contingent universal claims about the way the world is. If influential biologists and philosophers are right, there is nothing more to biological theorizing than these mathematical models. Thus, John Beatty (1981) writes, "[A biological] theory is not comprised of laws of nature.

Rather, a theory is just the specification of a kind of system—more a definition than an empirical claim." (p. 410). Models do not state empirical regularities, do not describe the behavior of phenomena; rather, they define a system. Here, Beatty follows Richard Lewontin (1980): in biology, "theory should not be an attempt to say how the world is. Rather, it is an attempt to construct the logical relations that arise from various assumptions about the world. It is an 'as if' set of conditional statements" (p. 58). Like Euclidean geometry, theoretical models in biology are not empirical claims; they are deductive systems. The only empirical questions are about where and when one or another of them is realized. This is a view of theories that has been expressed before, by instrumentalist philosophers of science. Despite its known limitations as a general account of scientific theorizing, we may now have reason to believe that it is more adequate a conception of biological theorizing than of theorizing in physical science. At least we can see why in biology it may be impossible to say how the world is nomologically speaking and it may be important to seek a second best alternative to this end.

Unlike physical science, where models are presumed to be way stations toward physical truths about the way the world works, there can be no such expectation in biology. Models of varying degrees of applicability for limited ranges of phenomena are the most we can hope for. A sequence of models cannot expect to move toward complete predictive accuracy, not even explanatory unification. The reasons should be clear in the impossibility of biological laws.

Consider the set of models that characterize population biology, models that begin with a simple two-locus model that reflects Mendel's "laws" of independent assortment and segregation, but which must be continually complicated, because Mendel's two laws are so riddled with exceptions that it is not worth revising them to accommodate their exceptions. These models introduce more and more loci, probabilities, recombination rates, mutation rates, population size, and so forth. And some of these models are extremely general, providing a handle on a wide range of different empirical cases; some are highly specific and enable us to deal with only a very specific organism in one ecosystem.

But what could the theory be like that underlies and systematizes these Mendelian models? First, suppose that since the models' predicates are all functional, the theory will be expressed in functional terms as well. But we know already that any theory so expressed will itself not provide the kind of exceptionless generalizations that would systematize the models in question; that will explain when they obtain and when they do not obtain. For this theory about functionally described objects itself will be exception-ridden for the same reasons the models of functionally described objects are.

Could a theory expressed in nonfunctional vocabulary systematize these models, explain when they work and when they do not? Such a theory cannot logically or even biologically be excluded, but its existence would vindicate a sort of reductionism that few biologists any longer anticipate. A nonfunctional theory, that is, a theory whose kind terms are structural or physical, would be nothing more or less than a body of physical and/or chemical laws. For such a theory to systematize biological models, the functional kinds of biology would have to be analyzed into complete lists of the diverse constituent physical components that can variously realize or implement these functional kinds. Selection has proceeded so far in our biocosm that any such analysis would be complex and unwieldy beyond our powers to employ it in real time prediction or humanly intelligible explanation. If such a physical theory reducing functional biology is a possibility, it is one of only Platonic interest. We know it exists because in the end the biological is nothing but physical, and the physical world is governed by strict laws. But this is cold comfort to creatures like us. For the functional kinds in which we can interest ourselves, even at the level of the molecular, cannot be linked up in this way with physical types, else there would be biological laws we could discover. Perhaps there are some functional individuations at levels of organization low enough so that nature does not have a chance to select multiple structures with similar effects. At this level, a physical theory will be able to explain by reduction, functional models. But for creatures of our cognitive and computational powers and our predictive and explanatory interests, the understanding that biological models provide is not epistemically underwritten by the possibility of more fundamental theory. And this raises the most serious epistemological issues for biology.

Many philosophers of science will be unhappy with this relativization of biology to our cognitive limits and interests. They will argue that this conclusion simply reflects a fundamental misconception of biology on the inappropriate model of physical science. According to some of these philosophers, biology differs from chemistry and physics not because of its limits but because of its aims. Unlike physics or chemistry, biology is a historical discipline, one that embodies neither laws nor aims at predictions any more than human history does. This approach stems in part from Darwin's own conception of the theory of natural selection as being in large measure a historical account of events on this planet and therefore not a body of empirical generalizations that if true, are true everywhere and always. Contemporary versions of this approach surrender Darwinian pretensions to nomological status for the theory of natural selection and defend its epistemic integrity as an historical account of the diversity and adaptedness of contemporary flora and fauna.

Such treatments of biology as history may even take comfort from what Clegg (Chapter 2, this volume) tells us about the aims of population genetics:

> In more recent years, the focus has moved from manipulative experiments to historical inference. The interest in historical inference has been stimulated by molecular data that permit direct comparisons of gene sequences among species. These data are then used to infer phylogenies or to infer the evolutionary history of particular genes and gene families. (pp. 36–37)

Clegg recognizes limits to historical knowledge; limits as sharp in human history as in biological history, and he goes on to say:

> There are strict limits to historical inference. It is not possible to "know" that the particular history estimated is the true history, although we may be able to say with a given probability that the true history lies within some ensemble of histories. . . .
>
> Moreover, the goal of historical inference is only approximately attainable, . . . (p. 37)

There are, however, two questions raised by the elevation of history as a central focus of evolutionary biology. The first is what kind of understanding biology can secure if its explanations are historical and not nomological; the second is the question of why we want historical knowledge in the first place.

As technology advanced, molecular biology increasingly was able to answer questions about histories, or at least increasingly to narrow the range of histories within which the true history must lie. Clegg suggests that it may now have reached its limits because nucleotide substitutions are the basement level of evolutionary change: "Because [DNA sequence] data are the most elemental, no further technology can be expected to provide enhanced genetic resolution" (p. 39). But let us suppose that we really can identify true histories, either because there is only one genealogy consistent with the data or because some other breakthrough identifies the one true history among its competing alternatives. Once we have the true sequence, what could we learn about why that sequence of changes emerged, and what might we learn about the general mechanism of adaptation and speciation as a whole?

Studies in the philosophy of history inspired by an empiricist philosophy of science some 50 years ago suggested that to explain why a particular sequence of events obtained, we need causal laws that link events in that sequence. Otherwise our explanation of why one event leads to another is mere chronicle, description, lacking in justification for its causal claims. But in human history there seem to be no laws of the sort needed to underwrite causal claims. Empiricist philosophers of science replied that historical explanations were incomplete "explanation sketches" that presupposed as yet undiscovered "implicit" laws. But few historians accepted this relegation of their explanations to incompleteness. The result was a search for other schemes of explanation, ones different from those of empirical science. These searches proposed that historical understanding is radically different from the sort of understanding that natural science provides. It provides empathetic understanding, or explains by redescription, or links explaining and explained events nonempirically.

Return now from human history to biology. The ancient empiricist account of historical understanding as law governed can hardly avail us in the attempt to understand biological understanding. For we have seen that there are no biological laws of the sort needed to informatively link even nucleotide substitutions, let alone events in a chain of adaptational evolution. Even the notion of an explanation sketch dependent on implicit laws will not work. The only laws to be found in biology are the principles of the theory of natural selection. By themselves these are too general to inform the detailed sequences we uncover. Showing that a sequence's data satisfy a certain model leaves open the causal question of why they do so as opposed to satisfying another model, as it does the question of when the sequence begins to or ceases to satisfy the model. These questions are not answered by the bland assurance that the values of variables that would make the model inapplicable are too low to do so. For the question is, "Why they are too low?" As we seek casual explanation in biology, the stratagems of a noncausal approach to explanation will not avail either.

When we explain a genealogy of nucleotide changes as the result of selection and not neutral drift, we are committed to the operation of a law or laws, which together with the values of variables and parameters determine the end point of the genealogy given the starting point. On the one hand, these laws will have to be more specific and more detailed than the principles of natural selection alone; on the other hand, they cannot be, for there are no laws in biology at lower levels of generality than those that state that variation and selection make for adaptational change.

The question remains: Once we have tied down the actual genealogy of a gene, a gene complex, a line of descent, or even the trajectory of a population through an epoch of geological change to a single track, what kind of understanding can we secure for it, if not a causal one?

The inescapability of the demands for causal explanations, and thus laws, becomes clear when we consider the second of the two questions posed above. Why do we want to know the actual histories of biological processes on this planet? Why are we interested in history? Is our interest intrinsic, like that of many human historians who seek to know what happened without any pretensions to why it had to happen that way? Or is the biologist's interest in history extrinsic and instrumental?

For Clegg, the interest in history is clearly enough extrinsic:

> Our strategy in exploring the limits to knowledge is to study the particular rather than the general. We will focus our attention on very specific data sets and we will use computer simulation as a means of further exploring the fit of particular historical scenarios to those data sets. Ours is a reductionist approach. We hope to dissect particular data sets in increasing detail and then to argue by induction that the resulting boundaries are general. (from an earlier draft of Chapter 2, this volume)

The interest in history evinced here is clearly instrumental. Clegg seeks to frame a general theory and test it through the study of data about the past. This is no surprise, given his interest in application to policy (i.e., to the future). But this sort of interest in history as a test bed for theory presupposes that there are general hypotheses narrow enough to be meaningfully tested by historical data. It presupposes not just that the data may confirm more strongly to one description of the actual sequence than others. For such descriptions are not generalizations. And it is general laws that are the objects of Clegg's inductive interests in history.

The role of history and the absence of laws have suggested to some biologists and philosophers that biology is a historical discipline in a sense different from Clegg's. It suggests that we seek to know the sequence of past events for its own intrinsic interest, not because as a retrodiction it can serve to test general biological hypotheses. For these biologists and philosophers the answer to the question, "Why do we want historical knowledge to begin with" is that we want it because, like human history, biological history sets the limits to the rest of biology.

Here is how Elliot Sober (1994) sees matters:

> The two main propositions in Darwin's theory of evolution are both historical hypotheses. . . . The ideas that all life is related and that natural selection is the principle cause of life's diversity are claims about a particular object (terrestrial life) and about how it came to exhibit its present characteristics. . . .
>
> Evolutionary theory is related to the rest of biology in the way the study of history is related to much of the social sciences. Economists and sociologists are interested in describing how a given society currently works. For example, they might study the post World War II United States. Social scientists will show how causes and effects are related within the society. But certain facts about that society. But certain facts about that society—for instance its configuration right after World War II—will be taken as given. The historian focuses on thses elements and traces them further into the past.
>
> Different social sciences often describe their objects on different scales. Individual psychology connects causes and effects that exist within an organism's own life span. Sociology and economics encompass longer reaches of time.

> And history often works in an even larger time frame. This intellectual division of labor is not entirely dissimilar to that found among physiology, ecology, and evolutionary theory. (p. 7)

So, evolutionary theory is to the rest of biology as history is to the social sciences. History is required for complete understanding in biology because biological theories can only provide an account of processes within time periods of varying lengths and not across several or all time periods. Why might this be true? This thesis will be true where the fundamental generalizations of a discipline are restricted in their force to a limited time period. History links these limited-time disciplines by tracing out the conditions that make for changes in the generalizations operative at each period. Some "historicist" philosophers of history, like Marx or Spengler, argue for time limits on explanatory generalizations by appealing to an ineluctable succession of historical epochs, each of which is causally required for every epoch that follows, no matter how long after they have occurred. Others have held that the best we can do is uncover epoch-limited generalizations and identify the historical incidents that bring them into and withdraw their force.

As a historical science, the absence of exceptionless generalizations in biology will be no more surprising than their absence is in history. For such generalizations seek to transcend temporal limits, and it is only within these limits apparently that historical generalizations are possible.

Leaving aside the broad and controversial questions about this sort of historicism in general, it is clear that many social scientists, especially sociological and economic theorists, will deny that the fundamental theories of their disciplines are related to history in the way Sober (1994) describes. Consider economic theory. Though economic historians are interested in describing and explaining particular economic phenomena in the past, the economic theory they employ begins with generalizations about rational choice that they treat as truths about human action everywhere and always. The theory is supposed to obtain in postwar America and in 7th-century Java. Similarly, sociological theory seeks to identify universal social forces that explain commonalities among societies due to adaptation and differences between them owing to adaptation to historically different local geographical, meteorological, agricultural, and other conditions.

For these sociological or economic theorists, the bearing of history is the reverse of Sober's picture. These disciplines claim, like physics, to identify fundamental explanatory laws; history *applies* theory to explain particular events that may test these theories.

Unlike Sober's view of the matter, on this view social science is more fundamental and history is derivative because it provides the theory. And if there is an analogy to biology it is that evolutionary theory is to the rest of biology as fundamental social theory is to history. This means that Sober is right about the rest of biology, but for the wrong reasons. Biology is a historical discipline, but not exactly because evolutionary theory is world-historical description and the rest of biology is temporally limited chronicle. This is almost right. The rest of biology is temporally limited history. But the main principles of Darwin's theory are not historical hypotheses. They are the only transtemporally exceptionless laws of biology. And it is the application of these laws to initial conditions that generates the functional kinds that make the rest of biology implicitly historical.

Evolutionary theory describes a mechanism—blind variation and natural selection—that operates everywhere and always throughout the universe. For evolution occurs whenever bits (tokens) of matter have become complex enough to foster their own replication so that selection for effects can take hold. And its force even dictates the functional individuations that terrestrial biologists employ and that thereby limit the explanatory and predictive power, and thus the scientific adequacy of their discipline.

In our little corner of the universe the ubiquitous process of selection for effects presumably began with hydrocarbons and nucleic and amino acids. That local fact explains the character of most of the subdisciplines of biology. Their explanations are "historically" limited by the initial distribution of matter on the earth and the levels of organization into which it has assembled itself. So, their generalizations are increasingly riddled with exceptions as evolution proceeds through time.

Thus, biology is a historical discipline because its detailed content is driven by variation and selection operating on initial conditions provided in the history of this planet, combined with our interest in functional individuation. This explains why biology cannot secure the sort of generality characteristic of the physical sciences. Its apparent generalizations are really spatiotemporally restricted statements about trends and the co-occurrence of finite sets of events, states, and processes. And there are no other generalizations about biological systems to be uncovered, at least none to be had that connect kinds under biological—that is, functional—descriptions. It is for this reason that there are limits to biological knowledge, limits more severe than the limits on knowledge in physical science, limits that stem not from the way natural selection generates the complexity that Clegg identifies as his sixth level of unknowability (see Chapter 2, this volume).

## REFERENCES

Beatty, J., 1981, What's wrong with the received view of evolutionary theory, in: *PSA 1980* (P. Asquith and R. Giere, eds.), pp. 397–439, Philosophy of Sciences Association, East Lansing, Michigan.

Fisher, R. A., 1957, *The Genetical Theory of Natural Selection*, 2nd ed., Dover, New York. (Originally published 1930)

Lewontin, R., 1980, Theoretical population genetics in the evolutionary synthesis, in: *The Evolutionary Synthesis* (E. Mayr and W. Provine, eds.), Harvard University Press, Cambridge, Massachusetts.

Sober, E., 1994, *The Philosophy of Biology*, Westview, Boulder, Colorado.

Wright, S., 1931, Evolution in Mendelian populations, *Genetics* **16:**97–159.

4

# Avoiding Paradigm-Based Limits to Knowledge of Evolution

WARD B. WATT

## INTRODUCTION

Since Darwin (1859) first proposed that evolution proceeds by natural selection, we have learned much about it. The founding of population genetic theory (summaries: Fisher, 1958; Haldane, 1932; Wright, 1931) showed the genetic feasibility of natural selection, removing a major objection to Darwin's theory (Provine, 1971), and led to extended study of population genetic phenomena (e.g., Nei, 1987; Hartl and Clark, 1989). The "Modern Synthesis" (Jepsen *et al.*, 1949; Mayr and Provine, 1980) brought paleontology and systematics together with population genetics to endorse Darwin's insights and, many thought, to lay the foundation of steady progress in understanding.

In some ways, this optimism may still be justified. We have seen major advances in the conceptualization of systematics (Michener and Sokal, 1957; Hennig, 1966; Mayr and Ashlock, 1969; Ridley, 1986), dramatic advances in paleontology (e.g., Whittington, 1985), and a major new source of evolutionary evidence in the availability of molecular data (e.g., Gillespie, 1991; Hillis *et al.*, 1996). However, concerns about the nature of our

---

WARD B. WATT • Department of Biological Sciences, Stanford University, Stanford, California 94305-5020; and Rocky Mountain Biological Laboratory, P.O. Box 519, Crested Butte, Colorado 81224.

*Evolutionary Biology, Volume 32*, edited by Michael T. Clegg *et al.*
Kluwer Academic / Plenum Publishers, New York, 2000.

understanding are common. The proper ways to study evolutionary adaptation (seen by Darwin as central to his theory), or even the utility of the adaptation concept at all, are now controversial. The continued utility of the Modern Synthesis has been questioned by some of its most distinguished intellectual offspring (e.g., Antonovics, 1987; Gould, 1980a,b). Clegg (Chapter 2, this volume) asks if population genetics has reached intrinsic limits to knowledge. Is there a resolution to this conflict of perceptions?

Traditional population genetic practice tries to infer the presence or absence of natural selection, and its nature if present, from data on population frequencies of alleles or genotypes, often without regard to biological processes going on within those populations. Though not impossible, this is often difficult (e.g., Gillespie, 1991). Clegg (Chapter 2, this volume) asks whether the difficulty may lie in the task of estimating basic parameters of population genetic theory. For example, he discusses the problems of estimating $\theta = 4N_e\mu$ ($N_e$ = effective population size, $\mu$ = mutation rate) from DNA sequence data under the assumption of sequence neutrality and the statistical difficulties of testing such data for the influence of selection (i.e., violation of the neutral assumption). He discusses the problems of inferring historical patterns of selection on flavonoid pigment biosynthesis in the morning glory (*Ipomoea purpurea*), using as evidence the sequence variation and divergence of genes coding for tissue-specialized versions of the first enzyme in the pigment pathway. Study of possibly adaptive variation in this system is said to be limited by the complexity of mapping from genes to phenotypes: The flavonoid pathway includes about a dozen enzyme steps and, *via* branch points along its length, leads to other biosyntheses as well as the "end product" anthocyanins.

Clegg's and Gillespie's discussions are unusual in at least considering the evolutionary study of biological mechanisms. By usual contrast, two recent evolutionary compendia contain little reference to the functional biology of organisms (Golding, 1994; Real, 1994). Many workers assume that population genetics should be both necessary and sufficient for the theoretical analysis of evolution (setting aside paleontology). However, Feder and Watt (1992) have argued that: (1) population genetic theory is necessary but not sufficient for evolutionary insight; and (2) the functional study of phenotype–environment interactions is equally necessary and is more able to decide which variants are adaptive, neutral, or constrained by various factors. This position, held by a few evolutionists for some time (e.g., Gillespie, 1991; Clark and Koehn, 1992; Watt, 1994), is ignored (e.g., Real, 1994) or even denied (e.g., Mayr, 1980, 1988) by others.

This chapter proposes that the evolutionary Modern Synthesis has led to an amechanistic worldview or paradigm (Kuhn, 1970) in evolutionary biology. Some of its common assumptions, notably that understanding of phenotypic mechanism is of minor relevance to evolutionary study or that population genetics is generally sufficient for evolutionary insight, pose serious limits to our knowledge of evolution. This is not to deny the accomplishments of the Modern Synthesis, but to say that studying biological mechanisms also is necessary to advance our understanding of evolution. We turn first to the applicability of the paradigm concept in biology.

## PARADIGMS AND WORLDVIEWS IN BIOLOGY

> Although I am fully convinced of the truth of the views given in this volume . . . , I by no means expect to convince experienced naturalists whose minds are stocked with a multitude of facts all viewed, during a long course of years, from a point of view directly opposite to mine. . . . [B]ut I look with confidence to the future,—to young and rising naturalists, who will be able to view both sides of the question with impartiality. (Darwin, 1859, Conclusion, p. 368)

Whorf (e.g., 1956) is often credited with formalizing the idea that human cognition is shaped by preexisting concepts or language, but the epigraph makes clear that Darwin was well aware of this. With respect to the practice of science, it has been most clearly formulated in Kuhn's (1970) concepts of paradigm and scientific revolution.

Briefly, Kuhn (1970) held that "normal science" is carried out by scientific communities, which, in any subdisciplinary area, work within a set of shared assumptions. These assumptions are not questioned in the course of normal science, and taken together (or embodied in a key example) constitute a "paradigm," or framework within which "puzzle-solving" studies normally go on. Scientific "revolutions" take place when paradigms break down: the common assumptions are overextended and can no longer accommodate new concepts or results. In revolutions of replacement, previous understandings are swept away; one such was Darwinian evolution's overthrow of the ideas of fixity of species and of special creation. In other instances, prior knowledge may still be thought valid after a revolution but finds a different place in new understanding (Kuhn, 1970, Chapter X). Such a revolution of reintegration is the incorporation of biochemistry, previously based mainly on molar concepts of chemical process, into the newer molecular biology, with its emphasis on genetic and molecular specificity for which molar chemistry sets boundary conditions (Judson, 1996). Kuhn recognized a continuum among such cases, though also stressing that different paradigms may entail incommensurability of concepts and measures derived from them.

Popper (e.g., retrospectively, 1994), Suppe (1977), and others criticized Kuhn for denying that scientists could rise above paradigmatic limitations of view and for arguing that science must normally operate in this self-limiting way (e.g., 1970, Chapters IX, XII). To them, Kuhn's views seemed to deny rational progress in scientific understanding or even the recourse to tests of evidence that is central to it. Kuhn (1970, "Postscript") clarified his position, considering how discussions between adherents of different paradigms resolve such clashes of view; some critics, such as Popper (1994), acknowledge this. Travis' and Collins' (1991) discussion of the warping of scientific peer review by scientists' failure to understand one another, due to lack of shared worldviews ("cognitive particularism"), does reemphasize that although scientists need not be dominated by paradigmatic limitations, they often *are* (cf. also Watt, 1995).

Mayr (1980, 1988) challenged application of the paradigm and scientific revolution concepts to evolutionary biology. He said that the Modern Synthesis arose from many sources, and viewed the study of evolution as a successive accumulation of thesis–antithesis–synthesis cycles. Others educated in the context of the Modern Synthesis often express similar views. But, as Hey (Chapter 5, this volume) shows, paradigmatic limits often are not perceived by those experiencing them. Kuhn discussed the "invisibility" of scientific revolutions at length (1970, Chapter XI). Assumptions about evolution that Mayr sees as evident truths (e.g., division of biology into the study of "proximate" and "ultimate" causes) may actually be paradigmatic limits on recognition of alternate views. Hey argues that the "Mendelizing" of Darwinian thought and the Modern Synthesis itself were each changes of evolutionary paradigm and suggests that other paradigm clashes may still be unresolved in the study of evolution (cf. also Cain and Provine, 1992).

We thus should take very seriously the application of Kuhn's paradigm and revolution concepts to evolutionary biology, especially if they are taken to include revolutions of reintegration as discussed above. If we are to rise above paradigmatic limitations, we need to recognize their possibilities realistically and ask how they may have shaped where we are now and how we got here.

## THE AMECHANISTIC PARADIGM IN EVOLUTIONARY STUDY

> . . . the clarification of the biochemical mechanism by which the genetic program is translated into the phenotype tells us absolutely nothing about the steps by

> which natural selection had built up the particular genetic program. (Mayr, 1980, pp. 9–10)

The origins of the amechanistic paradigm lie in Darwin's strong (but not exclusive) emphasis on the evolutionary importance of small variations. Fisher (1958, summary) built this idea into the base of the Modern Synthesis, deemphasizing the evolutionary study of phenotypic mechanism. First, Fisher's Fundamental Theorem of Natural Selection (1958, p. 37) made *additive* genetic variance central to his view of the operation of natural selection. Further, he argued that large heritable changes in phenotype are more likely to damage than to improve the phenotype (Fisher, 1958, pp. 42–44). This view is often called "micromutationism," which is a misnomer as it speaks to the *adaptive value* of variants, not the chance of their occurrence as mutants, in relation to the size of their impacts. Small changes in any structure or process are likely to have additive effects, supporting the Fundamental Theorem. Also, the effects of small changes are likely to be interchangeable, supporting the idea that total heritable variance in a character, not the specific nature of the variants, is adaptively important. This leads even to present-day expectations that selection may easily optimize phenotypes, given only enough heritable variance (e.g., Parker and Maynard Smith, 1990).

The Modern Synthesis reinforced the formal rather than mechanistic treatment of genotype–phenotype–environment relations. Most contributors to the Princeton Symposium dealt with varying genes as abstracted agents within a formalized phenotypic architecture (Jepsen *et al.*, 1949). Simpson (1949) reiterated Fisher's microvariationism, and later Mayr (1958) wrote (speculatively) about genes as "good mixers" or "jacks of all trades." Lerner (1954), while agreeing that "metric traits" might be polygenic, postulated general developmental canalization of phenotypes based on overdominant heterozygous genotypes of other, equally unspecified genes. These views reflected ignorance of genotype–phenotype relations, but more was involved: Evolutionists wished to limit the scope of their tasks. Levins (1965, p. 372) praised Fisher's microvariationist focus on genetic variance as a summary variable; thus, he thought, microlevel genetic variants and their action mechanisms need not be studied in detail by evolutionists working at the macrolevel. MacArthur (1965) said that population biologists should be concerned solely with "strategy"—Levins' macrolevel issues—in ecology and evolution, in contrast to cell biologists who were interested only in mechanism. The use of quantitative genetic theory, with a primary focus on the analysis and manipulation of additive genetic variance, as an evolutionary tool (e.g., Robertson, 1968; O'Donald,

1971; Lande, 1983; Lande and Arnold, 1983) has supported this amechanistic viewpoint. Adaptationists often assume polygenicity, hence quantitative genetic microvariationism and many-to-one genotype–phenotype relations for "important characters," to avoid studying the mechanisms of genotype–phenotype–environment interactions (e.g., Reeve and Sherman, 1993; Rose and Lauder, 1996a).

A strong expression of this amechanistic paradigm came in Mayr's splitting of biology between studies of molecular mechanisms and of evolutionary change:

> The functional biologist is interested in the phenotype and its development resulting from the translation of the genetic program within the framework of the environment of the specific individual. It is this interaction between the translation of the genetic program and the environment that we refer to as proximate causes. The evolutionist is interested in the origin of the genotype, in the historical reasons of antecedent adaptation and speciation responsible for the particular genetic program that exists. This analysis deals with ultimate causes. (Mayr, 1980, p. 9)

This argument may have been a reaction to ill-judged claims by overenthused molecular biologists, naively proposing to reduce all of evolutionary biology to chemistry. Nonetheless, it has worked against the needed recognition that study of heritable phenotypic mechanisms and their environmental interactions is crucial to understanding the causes of fitness differences.

Indeed, while Mayr argued earlier (1963) that a readout of genotypic sequence (not then feasible) would be the ultimate evolutionary evidence, he thereby fell into a serious trap: Contrary to this section's Mayrian epigraph, one cannot even perceive genotypic patterns accurately without knowing the phenotypic processes of genotypic expression. For instance, one cannot know why single mutants from a GGG sequence (in mRNA) to AGG or GGA might be lethal or selectively neutral, respectively, unless one knows the *phenotypic* facts of translation: AGG codes for arginine, whereas both GGG and GGA code for glycine. Genotype and phenotype are related not simply hierarchically but reciprocally because of such expression interactions and recursively because genotypes are changed over time by natural selection resulting from heritable phenotype–environment interactions (Feder and Watt, 1992). Thus, neither genotype nor phenotype can be understood in evolutionary terms without reference to the other. Further, while some amechanistic evolutionists hold that phenotypes are "open-ended" and less easily defined than genotypic patterns, this may simply reflect present ignorance of principles of evolved phenotypic organization (see below).

Adherents of the population biological, amechanist paradigm have certainly gained important insights into evolution and continue to do so. If simplifying assumptions have facilitated their success, well and good. What they have learned must be built on by broader viewpoints. Here, we argue only for the avoidance of the limitations inherent in their paradigm.

## CRISIS IN THE AMECHANISTIC EVOLUTIONARY PARADIGM

> In summary, the Synthesis placed restrictive notions on the conceptual richness and depth of evolutionary biology as a science. (Antonovics, 1987, p. 326)

If the amechanistic evolutionary paradigm was secure in its assumptions and intellectually productive without problems, there would be no need to consider its revision or discard. But it now appears that this paradigm constitutes a self-imposed limitation to knowledge for evolutionary biology. We next consider its difficulties.

### Intellectual Confinement by the Amechanistic Paradigm

> Most studies of natural selection contain . . . symptoms of a fundamental . . . lack of interest in the organisms . . . and little interest in why natural selection can occur. (Endler, 1986, p. 163)

Endler (1986) found much progress in the study of natural selection, but also noted many limitations. Estimates of lifetime fitness were rare. There was little effort to build on strength by studying "model" test systems, so at most a few traits were understood in any one taxon. Many workers were content to show the mere occurrence of natural selection, avoiding the question why variation in some trait(s) actually resulted in fitness differences among the variants. Evolutionists, he argued, needed to study functional variation in phenotypes and/or their interactions with ecological variables.

Antonovics (1987) reinforced these concerns. "Why . . ." he wrote, "were we still excited [merely] by demonstrating natural selection?" (p. 324). This might have been understandable in 1949, he argued, but by the late 1960s, let alone 1987, deeper studies were to be expected. He deplored the synthesis indifference to mechanistic, quantitative study of evolutionary processes or to rigorous tests of hypotheses and of *de novo* predictions from initial *post hoc* analysis and criticized overreliance on "the comparative method" as a substitute for experimental work in natural populations.

Orr and Coyne (1992) questioned whether microvariationism truly applies to adaptive differences among species. They found that positive evidence is scarce and that usual methods have little power to detect deviations from polygenic or microvariationist assumptions. For example, bristle number in *Drosophila*, an apparently polygenic trait, is actually oligogenic—controlled by a few genes at which alleles of major effect segregate—but focused experiments were needed to show this (Mackay, 1995).

Despite such expressions of uneasiness, many evolutionists have been slow to accept a need for more mechanistic studies. As a result of continuing reliance on amechanistic views, we now appear to be experiencing an intensifying paradigm crisis.

## Evidence for Paradigm Crisis

> Normal science . . . is predicated on the assumption that the scientific community knows what the world is like. (Kuhn, 1970, p. 5)

> It ain't necessarily so. (Gershwin, 1935)

Kuhn identified one clear symptom of paradigm crisis as the emergence of *empirical anomalies*: results that are not predicted, or otherwise inexplicable, by assumptions or principles accepted in a field. In current evolutionary biology, a major empirical anomaly is the repeated finding of variants at single genes having large, adaptive phenotypic effects with strong fitness consequences in the wild, affecting "complex characters" such as locomotion, osmoregulation, antipredator camouflage, and so forth (Endler, 1986). Such cases involving allozymes (e.g., Powers *et al.*, 1991; Watt, 1985a, 1994; Eanes, 1999), gene product expression levels (e.g., Crawford and Powers, 1989; Osborne *et al.*, 1997), color polymorphisms (e.g., Kettlewell, 1955; Nielsen and Watt, 1998), to mention only a few, are seen in diverse taxa from bacteria to mammals. They are unexpected by "neutralists" and "optimizing selectionists" alike, but are so frequent that they cannot be set aside as "exceptions." They falsify the generality of Fisher's microvariationism, suggesting instead that adaptive variants show a wide range of magnitudes of phenotypic effect.

Another symptom of paradigm crisis is that *clearly formulatable questions are excluded* by a field. MacArthur (1965) expressed disinterest in mechanistic questions on behalf of "population biologists," and this has been echoed repeatedly, as in Mayr's "proximate versus ultimate cause" dichotomy. Ewens and Feldman (1976) complained that even within "population biology," population geneticists would not "use sufficient ecology" to discriminate neutral from selected variation. The feasibility of functional study of adaptive variation at identified genes in the wild often has been shown, as noted above, but many evolutionists still ignore it or consider it to be exceptional or very difficult. This may result from fear of becoming enmeshed in intractable phenotypic complexity, which was of concern to the Modern Synthesis, but that itself shows the self-limitation to knowledge imposed by paradigmatic assumptions.

A third symptom of paradigm crisis is the *inability to resolve well-formulated questions* within limits of the paradigm's assumptions. A prime example is the unproductive (Watt, 1985b, 1995) but still sometimes revisited (Skibinski *et al.*, 1993) neutralist–selectionist debate. A variety of causes have made this debate pointless, for example, the fact that diverse models that incorporate statistical exchangeability of alleles but otherwise make strictly opposed neutralist or selectionist assumptions nonetheless predict very similar distributions of various population genetic parameters (Rothman and Templeton, 1980; Gillespie, 1991, 1994). Thus, besides sampling limits on resolving power (Ewens and Feldman, 1976) [e.g., confidence limits around predicted plots so broad as to span half of the allowable "point space" (Watt, 1985b)], the opposed models simply cannot be distinguished by the *kinds* of evidence—abstract genetic frequency statistics—"allowed" by the terms of the debate.

In summary, where evolutionists find seemingly anomalous results, are unable to recognize certain questions as pertinent, or lack answers to well-posed questions as in the case of *Ipomoea* (Chapter 2, this volume), the cause may not be "unknowability," but rather an intellectual crisis arising from the self-limitation of the amechanistic paradigm, which neglects the nature of adaptation, and thus cannot address the causes of fitness differences or equivalences among genotypes. Adding tools of comparative biochemistry, physiological ecology, and so forth to the powerful molecular analyses deployed by Clegg and colleagues might allow direct study of adaptive molecular evolution in *Ipomoea*. Population genetics, for all its power within its own domain, has no language in which to ask, for example, why variants have certain fitness values in certain habitats (Feder and Watt, 1992). To escape these amechanistic limits to knowledge, we must refocus on the concept of adaptation.

## ADAPTATION, NOT ADAPTATIONISM: AVOIDING ANOTHER LIMITING PARADIGM

Darwin did not use modern terms, but he saw clearly that adaptation is expressed in the phenotype–environment interaction that gives rise to natural selection. This interaction begins from some initial difference between character states in suitedness to environment ("aptation"; Gould and Vrba, 1982) and may be refined by successive rounds of selection. Except when environmental stringency is relaxed (Watt, 1986), the *result* of adaptedness is what we call Darwinian fitness: "the best chance of surviving and of procreating . . ." (Darwin, 1859, p. 63), an expectation of genotypes' reproductive success averaged over other variation (Nei, 1987; Feder and Watt, 1992). Adaptation, expressed in differential biological performance by heritable phenotypes in their habitats, is causally central to evolution by natural selection (Brandon, 1990; Feder and Watt, 1992).

The evolutionary process is divisible into four stages, which cycle recursively, generation after generation (Feder and Watt, 1992): *(1) genotypes → phenotypes*, where diverse adaptive mechanisms of molecular and organismal development interact with environmental surroundings; *(2) phenotypes → performances*, where alternative phenotypes exhibit different levels of adaptedness in the habitat-specific performance of immediate biological tasks; *(3) performances → fitnesses*, where the demographic consequences of adaptive performances are summed or integrated into absolute, then relative fitnesses; and *(4) fitnesses → genotypes*, the province of population genetics, where genotypic fitnesses interact with genetic drift and inbreeding through effective population size and breeding system to yield genotype frequencies for the next generation. This distinction between adaptive performance early in the evolutionary process, and the *consequent* appearance of fitness in the later stages, is essential to keep straight the logic of natural selection and to organize empirical studies of the process (Feder and Watt, 1992; Watt, 1994).

Clearly, then, mechanistic study of evolution implies seeking rigorous inquiry and generalization about adaptation. But, Gould and Lewontin (1979), challenging "evolutionary storytelling," attacked what they called "adaptationism," or the "Panglossian paradigm." Do we risk escaping the limits of amechanistic study, only to fall victim to adaptationist blinders on knowledge? We must ask how the adaptationist paradigm differs from what we propose here.

Rose and Lauder (1996b) identify adaptationism as ". . . a style of research . . . in which all features of organisms are viewed *a priori* as optimal features produced by natural selection specifically for current function" (p. 1). Mayr (1988) took a more conservative view: "A program of research intended to demonstrate the adaptedness of individuals and their characteristics . . ." (p. 149). Mayr agreed with Gould and Lewontin's critique of the "atomization" of phenotypes by "reductionist" studies that ignore holistic integration of organismal phenotypes. Otherwise, he defended the assumption of the need for historical–selectionist explanation of organismal features, as contrasted to earlier views such as development or atrophy through use or disuse, and so forth. Others since, for example, Parker and Maynard Smith (1990) or Reeve and Sherman (1993), hailed the assumption of adaptiveness as a virtue, while others have attacked it as a vice. Rose and Lauder (1996b), while noting the constrainedness or neutrality of some phenotypes, still speak of a "new adaptationism." This seems to be the "old" adaptationism with some corrections or exceptions: recognition that epistasis or gametic disequilibrium can impede optimizing selection (e.g., Karlin and Feldman, 1970), the need for rigorous phylogenetic analysis of interspecific comparisons (Felsenstein, 1985; Harvey and Pagel, 1991), and so on. But the question remains: Is it helpful, or legitimate, to associate the rigorous study of adaptive mechanisms with the assumption that adaptation is ubiquitous? Very clearly, no.

First, is it true in practice that adaptiveness is assumed? The usual null hypothesis in statistical testing is that there is no treatment effect. Thus, any statistical test of adaptive difference among character states assumes *ab initio* that there is *no* such difference; only if the null hypothesis can be rejected according to standard decision rules is an effect recognized. In short, the character states under study are initially assumed to be neutral, not adaptive. All the null models of population genetics itself, from the single-gene Hardy–Weinberg distribution onward in complexity, start with neutral assumptions. Tests for selection, whatever their discrimination power, test for deviations from neutrality. Any study of putative adaptive mechanisms that involves population genetic consequences of these mechanisms uses a neutral assumption as a point of departure. Whether or not adaptive departure is found does not change the nature of the starting point.

Mayr (1988) argued that we should test all possible adaptive explanations for phenotypes before considering the "unprovable" explanation of chance. But, this argument is entailed only by Mayr's historicist view of evolutionary studies:

> ... when one attempts to explain the features of something that is the product of evolution, one must attempt to reconstruct [its] evolutionary history ... This can only be done by inference. The most helpful procedure ... is to ask ... what is or might have been the selective advantage ... responsible for ... a particular feature. (p. 149)

If instead one can analyze a phenotype by testing among neutrality, constraint, or adaptedness with present-day experiments, Mayr's view is no longer entailed. A historical approach to living function sometimes may be indispensable, but it is not the only approach available to evolutionary biology.

Indeed, many complex features of biological structure or function arise from or are limited by constraining influences rather than either adaptiveness or the action of chance. Geometric or topological constraints may take a major hand, for example, in snail shells' form (Gould, 1989) or in the fractional-power scaling of metabolic processes with body mass (West *et al.*, 1997). Even where adaptive evolution is found, as in thermoregulatory color variation in butterflies (Watt, 1968), it may be limited by constraints of environmental variability or of interaction with other phenotypic "modules" (Wagner and Altenberg, 1996). In the former, genetic adaptation of butterflies to mean thermal habitat conditions is constrained by their need to withstand extremes of variation in those habitats (Kingsolver and Watt, 1984), while in the latter some characters that might evolve phenotypic plasticity (as a seasonal thermal adaptation) seem constrained from this by the negative impacts of such changes on the performance of the "flight motor" (Jacobs and Watt, 1994).

In short, adaptationism is not the same as a rigorous, experimental approach to the study of evolutionary mechanisms, which asks how, in functional terms, heritable phenotypic variants are neutral, constrained, or truly adaptive. Population-genetic theory plays its role in such analysis by setting boundary conditions, showing how genome structure and the genetic transmission process itself may constrain adaptive options, and clarifying the interaction of selection with randomizing drift or biasing inbreeding. Such combined study of both the early and the later parts of the evolutionary recursion may yield wholly new insight into how evolution works.

## ARE THERE PHILOSOPHICAL LIMITS TO KNOWLEDGE OF EVOLUTION?

Can specific evolutionary case studies be reintegrated into general, predictive understanding? Rosenberg (Chapter 3, this volume) argues no; that on philosophical grounds evolutionary biology is limited to post hoc, historical analysis, as follows:

1. Scientific principles have no valid explanatory or predictive power unless they
   a. are "strict laws"—laws that do not have "exceptions"—or,
   b. as "nonstrict" or *ceteris paribus* laws with exceptions, are supported by strict laws that can account independently for those exceptions.

2. There are no such biological laws, except natural selection itself, because
   a. evolution leads to individuation of biological entities by function, not structure, so functional classes of entities will be structurally heterogeneous and will not share "biologically interesting" structural features;
   b. therefore candidate biological laws will have "exceptions."
3. Existing, that is, population genetic, evolutionary models, subsumable under the theory of evolution by natural selection, are so complex that they cannot account realistically or practically for all the "exceptions" of other candidate laws.
4. Thus there is no lawfulness in biology having "the nomological force which scientific explanation requires," and biology, committed to functional individuation, "cannot secure the sort of generality characteristic of the physical sciences."

This is a thought-provoking argument; two responses are specially pertinent here.

First, do entities displaying *regularly* alternate behaviors in response to the same values of state variables, in biology or other sciences, really constitute "exceptions" to general laws? Even Newton's Laws of Motion suppose some qualification of their subjects (e.g., "bodies in motion"). Should one say that only qualifications of subject falling below some arbitrary threshold of complexity can support "laws"? If so, how is that threshold defined? As Mayr (1988) suggests, either principles governing pluralistic subjects, hence having alternate outcomes for regular subgroups of those subjects, are legitimate laws, or high-level generalizations of the "force which scientific explanation requires" may exist in some areas independently of laws governing simpler disciplines. The legitimacy of complexity in scientific laws has been defended by Suppe (1977), Tuomi (1981), or Lloyd (1994), who hold a hierarchy of laws and models to be essential in dealing with pluralistic phenomena. Low-level models, differing in detail because of applying to different regular subgroups of subjects, may yield distinct predictions and yet group together under a common general theory. Even such a "simple" physical concept as $F = ma$ (force = mass × acceleration) takes diverse forms, types of measurement, and so forth in diverse applications (Kuhn, 1970).

Second, even the "simple" interaction of chemical and physical principles with natural selection may generate far more predictive power than some suppose. For example, consider the propositions that "all mammals weigh more than 0.0001 g," or "all mammals weigh less than $5 \times 10^8$ kg (or some even more extreme number)". It is said that although these relate the structural property—weight—to "a" functional property—mammality (though class Mammalia is defined by more than one functional synapomorphy!)—they are not "biologically interesting laws" (Rosenberg, Chapter 3, this volume). However, the lack of interest results from inappropriate choice of weight, or mass, scaling. If, for example, the minimum mass for adult mammals is defined on a more sensible scale (Calder, 1984), we get an exceptionless, informative generalization about mammals, arising from the joint operation of laws that are strict by any standards: the laws of thermodynamics, which mandate laws of heat exchange and constraints on organismal energy budgets, and evolution by natural selection (Morowitz, 1978; Watt, 1986). These jointly set a lower size limit below which a mammal's continuous endothermy would lose so much energy as to preclude retaining enough for reproduction, or even for self-maintenance, hence lowered fitness (e.g., Calder, 1984; Watt, 1986). This size limit applies to other obligately endothermic animals—birds—but not to animals with other thermal strategies. It even can be empirically tested: We could artificially select on a very small mammal or bird species for even smaller adult size to test the prediction that such smaller adults would show poor adaptive physiological performance, hence decreased fitness, in their environments. Similar reasoning could test predictable loss of fitness in mammals that exceed size limits imposed by surface volume constraints on heat dissipation, by circulatory organization, and so on.

Thus, even those laws that philosophers already recognize as applying to biology allow general adaptive explanation and/or prediction (e.g., no adult mammals or birds smaller than *X* can evolve in the wild) in biology. It long has been clear that physics and chemistry, interacting with natural selection, can drive the adaptive evolution of phenotypes at levels ranging from organisms down to protein molecules (e.g., Monod *et al.*, 1963). This debate's outcome may simply echo Dobzhansky (1973): "Nothing in biology makes sense except in the light of evolution" (p. 125).

## PROSPECTS FOR EXTENDING THE LIMITS TO KNOWLEDGE OF EVOLUTION

### Examples of the Feasibility of Reintegrating Evolutionary Studies

If we are to have a mechanistic revolution in evolutionary biology, it should be one of genuine reintegration, not merely of clash with amechanistic views. Rather than repeating dialectic confrontation among extreme dogmas, as in the neutralist–selectionist controversy or in clashes between gradualists and punctuationists, evolutionists should use Platt's (1964) strategy of multiple simultaneous hypothesis testing, to avoid self-limitations of perspective (Watt, 1995). For example, if one asks how one can distinguish neutrality or constrainedness from adaptiveness among heritable variants, rather than "seeking to show" one or another outcome, one can recognize effective neutrality between alternatives even while these jointly differ from other variants of the same gene(s). The feasibility of such "bottom-up," mechanistic analysis of complex cases is seen, for example, in the study of natural variants of the *Colias* butterfly PGI (phosphoglucose isomerase) gene: genotypes 4/4 and 4/5 are equivalent in molecular function, organismal performance, and resulting fitness, while they differ from other PGI genotypes in performance and fitness components by several-fold, as predicted from molecular differences (e.g., Watt, 1985a,b, 1992).

Equally, evolutionary changes first observed at a macroscopic level can be traced down to their mechanistic adaptive origins, if any, while testing explicitly for nonadaptive alternatives. A striking case in point is the evolution by the bacterium *Escherichia coli* into a new liquid culture ecological niche over thousands of generations: Increase in the cells' level of adaptedness over time has been demonstrated by comparing cold-preserved ancestral stocks to their derived descendants in direct competitive studies (Lenski and Travisano, 1994). Nonadaptive explanations for changes in cell size were experimentally tested and rejected (Mongold and Lenski, 1996), and part of the adaptive physiological basis for these evolutionary changes has been traced to carbohydrate metabolism (Travisano and Lenski, 1996). Thus, mechanistic study of evolution can begin at any level of biological organization and proceed rigorously to encompass the others as needed.

Further, mechanistic analysis may be provoked by amechanistic findings and give more credibility and interpretive depth to such results. Consider the biasing of codon distributions within degenerate codon sets coding for single amino acids in protein-coding genes. While originally noted by "pure" sequence analysis, these findings were reinforced by the phenotypic matching of such biases to the intracellular concentrations of alternative tRNAs within the several degeneracy sets and by their recognition as a result of selection for message translation speed or precision in protein synthesis (Akashi, 1995; Akashi and Schaeffer, 1997; Bulmer, 1988). Such selection is quite weak on any one gene in the short run, but is effective over time. This has revived theoretical interest in the efficacy of weak selection pressures in general (Chapter 5, this volume).

## Conflict or Synergism? The Continued Utility of Subdisciplines

We do not claim here that evolutionary studies based within single subdisciplines are of no further use. Nothing could be further from the truth. For example, Aquadro (Chapter 7, this volume) and Hey (Chapter 5, this volume) argue cogently for the continuing vigor and productivity of new work within molecular population genetics, *sensu stricto*, and point out new and often surprising results that continue to be found in this area. For that matter, even now only much careful effort can take full advantage of the existing power of quantitative genetics, as noted above: highly focused experiments are needed to distinguish polygeny from oligogeny (Mackay, 1995). Yet the distinction, once made, is far-reaching: Microvariationism and its limited focus on "phenotypic evolution" are replaced by a view in which the identities and functions of "oligogenes" affecting bristle number become the productive focus of new research (Mackay, 1995).

However, these acknowledgments may not be taken as license for "business in isolation as usual." For example, molecular population genetics, *sensu stricto*, may well develop more sensitive kinds of testing for the action of natural selection on parts of genomes (Chapter 7, this volume), but as noted earlier, without help from mechanistic studies it cannot address the biological causes of such selection. The basic point here is that limitations to knowledge that seem fixed within subdisciplinary boundaries, may be easily removed when those boundaries are transcended.

## "Complexity Problems" in Adaptive Studies and Strategy to Address Them

Most mechanistic studies of adaptive variation discussed by, for example, Watt (1994) or Mitton (1997) resulted from a priori-designed investigation, not a string of unrelated, fortuitous discoveries. Nonetheless, some evolutionists still regard such cases as exceptions, achieved in spite of extraordinary difficulty. Dealing all at once with perceived complexities of genotypes → phenotypes, phenotypes → performances, and performances → fitnesses in the evolutionary recursion has daunted some workers. It may be difficult indeed to begin with the most phenotypically complex [e.g., Clark (Chapter 11, this volume) documents the problems of searching for genetic underpinnings of human diseases]. However, successes achieved thus far, in whatever system, do result from a clear strategy (Feder and Watt, 1992), not blind good fortune. Elements of this strategy include:

• Use well-known "model" systems to multiply the power of further studies. *Drosophila*, *Peromyscus*, *Clarkia*, *Fundulus*, *Anolis*, *Colias*, *Escherichia coli*, *Neurospora*, or *Mytilus* are all exemplary taxa whose well-understood biologies facilitate new case studies of evolutionary mechanisms (Perkins and Turner, 1988; Feder and Watt, 1992; Powell, 1997).

• Use knowledge of test species' specific biology to obviate limitations of those systems, such as generation time or difficulties of direct measurement of fitness components. Knowledge of *Colias*' reproductive biology led to a powerful design for study of male mating success among genotypes in these insects (Watt *et al.*, 1985). Coring trees for growth rings, or measuring rosette diameters in plants that grow for many years and flower once, together with genetic sampling, can allow evolutionists to assess genetic change over life cycles even in long-lived plants.

• Use more diverse conceptual or technical tools. Repeated use of familiar methods may involve less effort, but even that effort may be wasted if results are ambiguous (see Watt, 1985b for cases in point). At the same time, assimilation of new approaches must be thorough, or confusion may result. For example, Kacser and Burns (1981), recasting Wright's (1934) argument for physiological causes of dominance in terms of metabolic control theory, overstated the prevalence of dominance. Though this overstatement has been corrected, for example, by Watt (1985a) or Savageau and Sorribas (1989), some workers still use it to bolster claims that variation at single genes "should not" have large effects on metabolic or organismal phenotypes. Such claims ignore the fact that other work by the same authors (e.g., Kacser and Burns, 1979) or others (e.g., Mitton, 1997; Savageau and Sorribas, 1989; Watt 1985a,b, 1994) shows many cases in which metabolic effects of allelic variants at single genes are clear and large. Further, metabolic network theory, for steady-state or transient alike, provides sound bases in principle for such findings (e.g., Watt, 1985a, 1986; Savageau and Sorribas, 1989).

• Test multiple alternative hypotheses (Platt, 1964) to avoid unfruitful false dichotomies or dialectical squabbles, as noted earlier.

• Be aware that phenotypes must be internally organized (and interactive with their environments) in well-defined, often hierarchical, perhaps modular (Wagner and Altenberg, 1996) but empirically accessible ways, in contrast to claims of unanalyzable holism by adaptationists (Mayr, 1988) and antiadaptationists (Gould and Lewontin, 1979) alike. For example, biomolecules interact by the minimum number of steric contacts necessary for specificity, as expressed in Quastler's (1964, 1965) "signature principle." Equally, the complexity of branched metabolic pathways is controllable *in vivo* (as well as accessible to study) in part because of regulation of key metabolite concentrations at the joins of pathway branches. This isolates the branches as functionally independent modules, so that epistasis may occur among metabolic steps within branches but infrequently across branches (cf. Savageau and Sorribas, 1989). Finally, developmental "canalization" of phenotypic subsystems may insulate them from pleiotropic side effects of genetic variation in other subsystems, just as from environmental insults, as in the "crossveinless complex" of *Drosophila* (e.g., Milkman, 1961, 1965). In summary, phenotypes, while complex, are not "open-ended" or "indefinitely interconnected," as some assert. Biologists can address their organization, analyze phenotype–environment interactions, then resynthesize resulting discoveries.

A new a priori-designed investigation, addressing many of these issues, deals with heat-shock proteins in *Drosophila*. The ecological niche of larval *D. melanogaster* exposes both larvae and pupae to serious thermal stress in their natural microhabitat: fallen and rotting fruit (Feder *et al.*, 1997). There is genetic variation in Hsp 70 stress protein expression, and there are direct survival effects of this variation, predictable from interaction with thermal stress (Krebs and Feder, 1997). The trade-off of Hsp 70 thermal stress protection against cost of Hsp 70 production in benign habitats is a focus for further study. This work shows the feasibility of direct study of "complex" organism–environment interaction and the importance of using natural genetic variation to probe a phenotypic subsystem previously studied typologically (as Krebs and Feder point out).

## Broad Evolutionary Time Scales

Extension of the limits to knowledge need not be confined to the study of current evolutionary processes. We also can push back such limits in the study of long-time-scale evolution, at least setting sharper boundary conditions on inferences about the past. For example, C.S. Hickman (1988) has discussed diverse resources for studying the evolution of form in extinct animals: study of form–function patterns and processes in contemporary relatives, analogy to similar forms outside of biology, biomechanical modeling of fossils' anatomy in environmental context, and so forth.

New tools of molecular evolution also can illuminate such inference. For example, three (or more) hypotheses for the origins of insect wings have been debated for years. Molecular homology now has been found between a gene affecting insect wing development and one expressed in a plausibly wing-precursory part of the plesiomorphic arthropod biramous limb (Averoff and Cohen, 1997), at the very least sharpening the terms of this debate. Alternatively, adaptation at the molecular level can be analyzed in the historical context of phylogenetic reconstruction from independent molecular evidence (Golding and Dean, 1998). Such work will be aided by more study of the challenges of ancestral state reconstruction (e.g., Churchill, Chapter 6, this volume).

### The Possible Generality of Evolutionary Study

Contrary to philosophical arguments such as critiqued earlier, there is good evidence that evolution must be somewhat generalizable: for example, distantly related taxa may converge in many characters when they evolve to occupy similar niches (e.g., MacArthur, 1965). As a case in point, molecular phylogenetics shows convergence of diverse *Anolis* lizard lineages into a small, repeatable array of "ecomorphs" on the Greater Antilles (Losos *et al.*, 1998). Thus, as Hey (Chapter 5, this volume) emphasizes, a priori despair of generality or predictability in evolutionary study is neither heuristic nor justifiable. Our task is to discover how far adaptive evolution is predictable and what rules govern it. For this, we must move beyond the amechanistic paradigm. The increasing diversity of successful mechanistic evolutionary case studies shows that this is practical.

## CONCLUSIONS

Biologists have learned much about evolution since Darwin, but some are concerned about adequacy of present knowledge for further growth of understanding. Some have asked whether evolutionary biology is reaching limits to the "knowability" of our material. However, such limits may be self-imposed by would-be knowers, rather than by their materials. Current evolutionary biology has been sharply limited by an amechanistic paradigm, developed from the 1930s up to and through the Modern Synthesis, which has biased many scientists' thinking away from important approaches to evolution. This paradigm is critiqued as self-limiting. Increased evolutionary emphasis on biological mechanism requires renewed attention to adaptation, but not in the manner of the adaptationist paradigm, which is also critiqued as unhelpful and potentially misleading.

To break out of paradigmatic limits to evolutionary knowledge, we can use mechanistic approaches, whether experimental or observational, to distinguish among the adaptations, constraints, or neutrality (the fundamental null hypothesis) exhibited by variation in organismal phenotypes. Whether or not biology harbors general "laws" other than evolution by natural selection, the combination of natural selection with physical and chemical laws is already enough to yield considerable generality and predictive ability for mechanistic evolutionary biology. Success in this is not limited to fortuitous cases: Examples illustrate a feasible strategy for mechanistic evolutionary study of diverse, complex organism–environment interactions. This approach, with the aid of both molecular and morphological evidence, also may extend evolutionary inference through time to illuminate long-term as well as short-term problems. By reintegrating the successes of the modern synthesis with new and mechanistic approaches, ranging from molecular to ecological levels, we can energetically remove self-imposed limits to our knowledge of evolutionary biology.

ACKNOWLEDGMENTS

I thank Michael Clegg for this chance to consider the limits to knowledge of evolution. I thank Charles Aquadro, Carol Boggs, Frances Chew, Michael Clegg, Antony Dean, Paul Ehrlich, Martin Feder, Marcus Feldman, John Gillespie, Peter Godfrey-Smith, Dan Graur, John Harte, Mary Ellen Harte, Carole Hickman, Richard Lenski, Charles Remington, Joan Roughgarden, Clay Sassaman, Lee Snyder, and Steven Stefanides for discussions over the years. Work in our laboratory has been supported by the US Department of Energy (ER-03-93-61667), US National Institutes of Health (GM 26758), and several grants from the US National Science Foundation.

## REFERENCES

Akashi, H., 1995, Inferring weak selection from patterns of polymorphism and divergence at "silent" sites in *Drosophila* DNA, *Genetics* **139:**1067–1076.

Akashi, H., and Schaeffer, S. W., 1997, Natural selection and the frequency distributions of "silent" DNA polymorphism in *Drosophila*, *Genetics* **146:**295–307.

Antonovics, J., 1987, The evolutionary dys-synthesis: Which bottles for which wine? *Am. Natur.* **129:**321–331.

Averoff, M., and Cohen, S. M., 1997, Evolutionary origin of insect wings from ancestral gills, *Nature* **385:**627–630.

Brandon, R. N., 1990, *Adaptation and Environment*, Princeton University Press, Princeton, New Jersey.

Bulmer, M., 1988, Are codon usage patterns in unicellular organisms determined by mutation-selection balance? *J. Evol. Biol.* **1:**15–26.

Cain, A. J., and Provine, W., 1992, Genes and ecology in history, in: *Genes in Ecology* (R. J. Berry, T. J. Crawford, and G. M. Hewitt, eds.), pp. 3–28, Blackwell Scientific Publications, Oxford, England.

Calder, W. A., 1984, *Size, Function, and Life History*, Harvard University Press, Cambridge, Massachusetts.

Clark, A. G., and Koehn, R. K., 1992, Enzymes and adaptation, in: *Genes in Ecology* (R. J. Berry, T. J. Crawford, and G. M. Hewitt, eds.), pp. 193–228, Blackwell Scientific Publications, Oxford, England.

Crawford, D. L., and Powers, D. A., 1989, Molecular basis of evolutionary adaptation at the lactate dehydrogenase-B locus in the fish *Fundulus heteroclitus*, *Proc. Natl. Acad. Sci. USA* **86:**9365–9369.

Darwin, C., 1859, *The Origin of Species*, 6th ed., rev., 1872. New American Library, New York.

Dobzhansky, Th., 1973, Nothing in biology makes sense except in the light of evolution, *Amer. Biol. Teacher* **35:**125–129.

Eanes, W. F., 1999, Analysis of selection on enzyme polymorphisms. *Annu. Rev. Ecol. Syst.* **30:**301–326.

Endler, J. A., 1986, *Natural Selection in the Wild*, Princeton University Press, Princeton, New Jersey.

Ewens, W., and Feldman, M. W., 1976, The theoretical assessment of selective neutrality, in: *Population Genetics and Ecology* (S. Karlin and E. Nevo, eds.), pp. 303–337, Academic Press, New York.

Feder, M. E., and Watt, W. B., 1992, Functional biology of adaptation, in: *Genes in Ecology* (R. J. Berry, T. J. Crawford, and G. M. Hewitt, eds.), pp. 365–392, Blackwell Scientific Publications, Oxford, England.

Feder, M. E., Blair, N., and Figueras, H., 1997, Natural thermal stress and heat-shock protein expression in *Drosophila* larvae and pupae, *Funct. Ecol.* **11:**90–100.

Felsenstein, J., 1985, Phylogenies and the comparative method, *Am. Natur.* **125:**1–15.

Fisher, R. A., 1958, *The Genetical Theory of Natural Selection*, 2nd ed., rev., Dover, New York.

Gershwin, I., 1935, Libretto to George Gershwin, *Porgy and Bess.*

Gillespie, J. H., 1991, *The Causes of Molecular Evolution*, Oxford University Press, Oxford, England.

Gillespie, J. H., 1994, Substitution processes in molecular evolution, II. Exchangeable models from population genetics, *Evolution* **48:**1101–1113.

Golding, G. B. (ed.), 1994, *Non-neutral Evolution*, Chapman & Hall, New York.

Golding, G. B., and Dean, A. M., 1998, The structural basis of molecular adaptation, *Mol. Biol. Evol.* **15:**355–369.

Gould, S. J., 1980a, Is a new and general theory of evolution emerging? *Paleobiology* **6:**119–130.

Gould, S. J., 1980b, The evolutionary biology of constraint, *Daedalus* **109:**39–52.

Gould, S. J., 1989, A developmental constraint in *Cerion*, with comments on the definition and interpretation of constraint in evolution, *Evolution* **43:**516–539.

Gould, S. J., and Lewontin, R. C., 1979, The spandrels of San Marco and the Panglossian paradigm, *Proc. Roy. Soc. Lond. B* **205:**581–598.

Gould, S. J., and Vrba, E. S., 1982, Exaptation—A missing term in the science of form, *Paleobiology* **8:**4–15.

Haldane, J. B. S., 1932, *The Causes of Evolution*, Longmans, London.

Hartl, D. L., and Clark, A. G., 1989, *Principles of Population Genetics*, 2nd ed., Sinauer Associates, Sunderland, Massachusetts.

Harvey, P. H., and Pagel, M. D., 1991, *The Comparative Method in Evolutionary Biology*, Oxford University Press, Oxford, England.

Hennig, W., 1966, *Phylogenetic Systematics*, University of Illinois Press, Urbana, Illinois.

Hickman, C. S., 1988, Analysis of form and function in fossils, *Am. Zool.* **28:**775–793.
Hillis, D. M., Moritz, C., and Mable, B. K. (eds.), 1996, *Molecular Systematics*, 2nd ed., Sinauer Associates, Sunderland, Massachusetts.
Jacobs, M. D., and Watt, W. B., 1994, Seasonal adaptation *vs* physiological constraint: Photoperiod, thermoregulation, and flight in *Colias* butterflies, *Func. Ecol.* **8:**366–376.
Jepsen, G. L., Simpson, G. G., and Mayr, E. (eds.), 1949, *Genetics, Palaeontology, and Evolution*, Princeton University Press, Princeton, New Jersey.
Judson, H. F., 1996, *The Eighth Day of Creation*, 2nd ed., Cold Spring Harbor Laboratory Press, Plainview, New York.
Kacser, H., and Burns, J. A., 1979, Molecular democracy: Who shares the controls? *Biochem. Soc. Trans.* **7:**1149–1160.
Kacser, H., and Burns, J. A., 1981, The molecular basis of dominance, *Genetics* **97:**639–666.
Karlin, S., and Feldman, M. W., 1970, Linkage and selection: Two-locus symmetric viability model, *Theor. Pop. Biol.* **1:**39–71.
Kettlewell, H. B. D., 1955, Selection experiments on industrial melanism in the Lepidoptera, *Heredity* **9:**323–342.
Kingsolver, J. G., and Watt, W. B., 1984, Mechanistic constraints and optimality models: Thermoregulatory strategies in *Colias* butterflies, *Ecology* **65:**1835–1839.
Krebs, R. A., and Feder, M. E., 1997, Natural variation in the expression of the heat-shock protein *hsp70* in a population of *Drosophila melanogaster* and its correlation with tolerance of ecologically relevant thermal stress, *Evolution* **50:**173–179.
Kuhn, T. S., 1970, *The Structure of Scientific Revolutions*, 2nd ed., University of Chicago Press, Chicago.
Lande, R., 1983, The response to selection on major and minor mutations affecting a metrical trait, *Heredity* **50:**47–65.
Lande, R., and Arnold, S. J., 1983, The measurement of selection on correlated characters, *Evolution* **37:**1210–1226.
Lenski, R. E., and Travisano, M., 1994, Dynamics of adaptation and diversification: A 10,000-generation experiment with bacterial populations, *Proc. Natl. Acad. Sci. USA* **91:**6808–6814.
Lerner, I. M., 1954, *Genetic Homeostasis*, Oliver & Boyd, Edinburgh, England.
Levins, R., 1965, Genetic consequences of natural selection, in: *Theoretical and Mathematical Biology* (T. H. Waterman and H. J. Morowitz, eds.), pp. 371–387, Blaisdell, New York.
Lloyd, E. A., 1994, *The Structure and Confirmation of Evolutionary Theory*, 2nd ed., Princeton University Press, Princeton, New Jersey.
Losos, J. B., Jackmann, T. R., Larson, A., de Queiroz, K., and Rodriguez-Schettino, L., 1998, Contingency and determinism in replicated adaptive radiations of island lizards, *Science* **279:**2115–2118.
MacArthur, R. H., 1965, Ecological consequences of natural selection, in: *Theoretical and Mathematical Biology* (T. H. Waterman and H. J. Morowitz, eds.), pp. 388–397, Blaisdell, New York.
Mackay, T. F. C., 1995, The genetic basis of quantitative variation: numbers of sensory bristles of *Drosophila melanogaster* as a model system, *Trends Genet.* **11:**464–470.
Mayr, E., 1958, Change of genetic environment and evolution, in: *Evolution as a Process* (J. Huxley, A. C. Hardy, and E. B. Ford, eds.), 2nd ed., pp. 188–213, Allen & Unwin, London.
Mayr, E., 1963, *Animal Species and Evolution*, Harvard University Press, Cambridge, Massachusetts.
Mayr, E., 1980, Some thoughts on the history of the evolutionary synthesis, in: *The Evolutionary Synthesis* (E. Mayr and W. B. Provine, eds.), pp. 1–48, Harvard University Press, Cambridge, Massachusetts.

Mayr, E., 1988, *Toward a New Philosophy of Biology*, Harvard University Press, Cambridge, Massachusetts.
Mayr, E., and Ashlock, P. D., 1969, *Principles of Systematic Zoology*, 2nd ed., McGraw-Hill, New York, New York.
Mayr, E., and Provine, W. B. (eds.), 1980, *The Evolutionary Synthesis*, Harvard University Press, Cambridge, Massachusetts.
Michener, C. D., and Sokal, R. R., 1957, A quantitative approach to a problem in classification, *Evolution* **11:**130–162.
Milkman, R. D., 1961, The genetic basis of natural variation. III. Developmental lability and evolutionary potential, *Genetics* **46:**25–38.
Milkman, R. D., 1965, The genetic basis of natural variation. VII. The individuality of polygenic combinations in *Drosophila*, *Genetics* **52:**789–799.
Mitton, J. B., 1997, *Selection in Natural Populations*, Oxford University Press, New York.
Mongold, J. A., and Lenski, R. E., 1996, Experimental rejection of a nonadaptive explanation for increased cell size in *Escherichia coli*, *J. Bact.* **178:**5333–5334.
Monod, J., Changeux, J.-P., and Jacob, F., 1963, Allosteric proteins and cellular control systems, *J. Mol. Biol.* **6:**306–309.
Morowitz, H. J., 1978, *Foundations of Bioenergetics*, Academic Press, New York.
Nei, M., 1987, *Molecular Evolutionary Genetics*, Columbia University Press, New York.
Nielsen, M. G., and Watt, W. B., 1998, Behavioral fitness components in the "alba" polymorphism of *Colias* (Lepidoptera, Pieridae): Adult time budget analysis, *Func. Ecol.* **12:**149–158.
O'Donald, P., 1971, Natural selection for quantitative characters, *Heredity* **27:**137–153.
Orr, H. A., and Coyne, J. A., 1992, The genetics of adaptation: a reassessment, *Am. Natur.* **140:**725–742.
Osborne, K. A., Robichon, A., Burgess, E., Butland, S., Shaw, R. A., Coulthard, A., Pereira, H. S., Greenspan, R. J., and Sokolowski, M. B., 1997, Natural behavior polymorphism due to a cGMP-dependent protein kinase of *Drosophila*, *Science* **277:**834–836.
Parker, G. A., and Maynard Smith, J., 1990, Optimality theory in evolutionary biology, *Nature* **348:**27–33.
Perkins, D. D., and Turner, B. C., 1988, *Neurospora* from natural populations: Toward the population biology of a haploid eukaryote, *Exp. Mycol.* **12:**91–131.
Platt, J. R., 1964, Strong inference, *Science* **146:**347–353.
Popper, K. R., 1994, *The Myth of the Framework*, Routledge, London.
Powell, J. R., 1997, *Progress and Prospects in Evolutionary Biology: The Drosophila Model*, Oxford University Press, Oxford, England.
Powers, D. A., Lauerman, T., Crawford, D., and DiMichele, L., 1991, Genetic mechanisms for adapting to a changing environment, *Annu. Rev. Genet.* **25:**629–659.
Provine, W. B., 1971, *The Origins of Theoretical Population Genetics*, University of Chicago Press, Chicago.
Quastler, H., 1964, *The Emergence of Biological Organization*, Yale University Press, New Haven, Connecticut.
Quastler, H., 1965, General principles of systems analysis, in: *Theoretical and Mathematical Biology* (T. H. Waterman and H. J. Morowitz, eds.), pp. 313–333, Blaisdell, New York.
Real, L. A. (ed.), 1994, *Ecological Genetics*, Princeton University Press, Princeton, New Jersey.
Reeve, H. K., and Sherman, P. W., 1993, Adaptation and the goals of evolutionary research, *Q. Rev. Biol.* **68:**1–32.
Ridley, M., 1986, *Evolution and Classification*, Longman's, Harlow, England.
Robertson, A., 1968, The spectrum of genetic variation, in: *Population Biology and Evolution* (R. C. Lewontin, ed.), pp. 5–16, Syracuse University Press, Syracuse, New York.

Rose, M. R., and Lauder, G. V. (eds.), 1996a, *Adaptation*, Academic Press, New York.
Rose, M. R., and Lauder, G. V., 1996b, Post-spandrel adaptationism, in: *Adaptation* (M. R. Rose and G. V. Lauder, eds.), pp. 1–8, Academic Press, New York.
Rothman, E., and Templeton, A. M., 1980, A class of models of selectively neutral alleles, *Theor. Pop. Biol.* **18:**135–150.
Savageau, M. A., and Sorribas, A., 1989, Constraints among molecular and systemic properties: implications for physiological genetics, *J. Theor. Biol.* **141:**93–115.
Simpson, G. G., 1949, *The Meaning of Evolution*, Yale University Press, New Haven, Connecticut.
Skibinski, D. O. F., Woodwark, M., and Ward, R. D., 1993, A quantitative test of the neutral theory using pooled allozyme data, *Genetics* **135:**233–248.
Suppe, F., 1977, Afterword, in: *The Structure of Scientific Theories* (F. Suppe, ed.), pp. 617–730, University of Illinois Press, Urbana, Illinois.
Travis, G. D. L., and Collins, H. M., 1991, New light on old boys: Cognitive and institutional particularism in the peer review system, *Sci. Tech. Hum. Values* **16:**322–341.
Travisano, M., and Lenski, R. E., 1996, Long-term experimental evolution in *Escherichia coli.* IV. Targets of selection and the specificity of adaptation, *Genetics* **143:**15–26.
Tuomi, J., 1981, Structure and dynamics of Darwinian evolutionary theory, *Syst. Zool.* **30:**22–31.
Wagner, G. P., and Altenberg, L., 1996, Complex adaptations and the evolution of evolvability, *Evolution* **50:**967–976.
Watt, W. B., 1968, Adaptive significance of pigment polymorphisms in *Colias* butterflies. I. Variation of melanin pigment in relation to thermoregulation, *Evolution* **22:**437–458.
Watt, W. B., 1985a, Bioenergetics and evolutionary genetics—Opportunities for new synthesis, *Am. Natur.* **125:**118–143.
Watt, W. B., 1985b, Allelic isozymes and the mechanistic study of evolution, *Isozymes: Curr. Top. Biol. Med. Res.* **12:**89–132.
Watt, W. B., 1986, Power and efficiency as fitness indices in metabolic organization, *Am. Natur.* **127:**629–653.
Watt, W. B., 1992, Eggs, enzymes, and evolution—Natural genetic variants change insect fecundity, *Proc. Natl. Acad. Sci. USA* **89:**10608–10612.
Watt, W. B., 1994, Allozymes in evolutionary genetics: self-imposed burden or extraordinary tool? *Genetics* **136:**11–16.
Watt, W. B., 1995, Allozymes in evolutionary genetics: beyond the twin pitfalls of "neutralism" and "selectionism," *Rev. Suisse de Zoologie* **102:**869–882.
Watt, W. B., Carter, P. A., and Blower, S. M., 1985, Adaptation at specific loci. IV. Differential mating success among glycolytic allozyme genotypes of *Colias* butterflies, *Genetics* **109:**157–175.
West, G. B., Brown, J. H., and Enquist, B. J., 1997, A general model for the origin of allometric scaling laws in biology, *Science* **276:**122–126.
Whittington, H. B., 1985, *The Burgess Shale*, Yale University Press, New Haven, Connecticut.
Whorf, B. L., 1956, *Language, Thought, and Reality* (J. B. Carroll, ed.), MIT Press, Cambridge, Massachusetts.
Wright, S., 1931, Evolution in Mendelian populations, *Genetics* **6:**97–159.
Wright, S., 1934, Physiological and evolutionary theories of dominance, *Am. Natur.* **34:**24–53.

5

# Anticipating Scientific Revolutions in Evolutionary Genetics

JODY HEY

## INTRODUCTION

In day-to-day research in evolutionary genetics, it often seems as though our knowledge is bounded. We steadily perceive at least two major limitations on our capacity to understand the mechanisms and history of evolution. First, knowledge seems limited by the nature of history. It is certainly not useful to pursue a historical record that does not exist, as may be the case for many kinds of histories. Not all events, evolutionary or otherwise, leave an imprint in DNA or other media; and of the imprints that are made, none are expected to last indefinitely. Second, for the special case of evolutionary genetic histories, knowledge seems limited by the irreducible nature of DNA sequences. It does an investigator little good to try and glean more information from DNA than is available in the DNA sequence. So far as we know, every "A" base (adenine), for example, is like every other, and the information in a DNA sequence is in quanta (it is digital, base 4). These everyday perceptions of limits to inquiry may seem reasonable, and so they may provide a starting point for accessing what kinds of questions are more feasible than others. Perhaps we cannot reveal all of evolutionary history, but maybe we can understand and assess the limits to our knowledge of this history.

---

JODY HEY • Department of Genetics, Nelson Biological Labs, Rutgers University, Piscataway, New Jersey 08854-8082.

*Evolutionary Biology, Volume 32*, edited by Michael T. Clegg *et al.*
Kluwer Academic / Plenum Publishers, New York, 2000.

Is anything wrong with this reasoning? Yes, something is wrong; it is incomplete and there are other issues that bear on research priorities. In this chapter I lay out a case for why a scientific focus on the limits of knowledge is a second-rate kind of science, at least for evolutionary genetics. Despite an accessible and reasonable motivation for a scientific focus on the limits of knowledge, I do not think that questions of that kind are sufficiently motivated for basic researchers to spend much time on them.

## QUESTIONS CLEAR AND QUESTIONS NOT

I think that the limit of our knowledge is a worthy subject when our evolutionary questions are strictly posed, such that the words in the question have precise and well-understood meanings. In some sciences, like applied mathematics, where words do have relatively precise and unchanging meanings, questions about the limits of knowledge are frequent. It is common to address a very difficult question not with an answer, which may not be practically attainable, but with a quantitative assessment of how difficult the question is. Even though the specific question may not be answered, some knowledge is gained in explaining why. This kind of metaknowledge still may carry a fair bit of intellectual satisfaction. However, if the question is important, such that it really mattered that it be resolved, then a concluding statement on the difficulty of the question, no matter how elegant or informative in other contexts, is not useful.

Some of our questions in evolutionary genetics are well posed in this way, and for them the meaning of the question is not a part of the puzzle and the words of the question are clear and invariant. These are often relatively small questions that are pressing for some practical or applied reason. For example, a corn breeder may ask an evolutionary geneticist, "When did maize diverge from its closest teosinte relatives?" The corn breeder needs a number, preferably with some assessment of confidence in that number. In our example, we suppose that the corn breeder does not care about the causes of uncertainty; it is the answer that is paramount. The clarity of the question and the need for the answer are pressing, even if one cause of uncertainty is that, to the evolutionary geneticist, words like "diverge" and "relatives" do not have precise meanings. In an evolutionary context "diverge" usually means to become different, but there are many ways to assess divergence. Also, divergence takes time, and so the question about "when" raises semantic difficulties, as it implies a distinct time point. Similarly, "relatives" in this example is used in reference to the relationship between two closely related species. But the meaning of "species" is one of our most famous and persistent ambiguities.

When the meaning of a question is clear to the person asking the question, and so long as the asker is not going to change those meanings, the question is strictly posed. In this case it is possible to consider the reasons why an answer might be difficult to obtain. Now consider instead that the same question is posed between two evolutionary geneticists. In this case, the question may very well go away or be changed considerably, as the discussion turns to the meanings of the words in the question. The uncertainty of the meanings of the words precludes discussion on the limits of knowledge. In this context the question is not strictly posed.

At any point in time researchers in a basic science are constrained by their concepts and their lexicon of the moment. Within those constraints it is possible to address the limits of knowledge. But a strong focus on the limits of knowledge is only interesting if we presume that our basic concepts and lexicon are fixed. We have seen that they may be fixed for practical reasons. The corn breeder has other things to worry about besides the details of uncertain evolutionary concepts, and so she assumes, for practical reasons, that the terms of her question are fixed. Another way that concepts and lexicon may be fixed is if our science has reached the truth, if no amount of additional evidence could cause our concepts and lexicon to undergo replacement or reduction. If somehow this did happen and we were aware of it, then further research on conceptual knowledge would stop, and probably the only purposeful work that remained for scientists would be to ask questions about the limits of our true knowledge. Of course it is impossible to know when we have the truth, so one might argue that we should never waste time thinking about the limits of knowledge. In practice, scientists necessarily have some level of confidence that their ideas correspond closely to universal truths. The higher the confidence, the more that questions about the limits of knowledge become appropriate, or at least seem to become appropriate.

Are the questions of evolutionary geneticists well posed? Do we have well-understood concepts behind the words we use, and do we have some confidence that those concepts will not be overturned? For many of the most interesting questions of broad scope, I do not think we are even close to these ideals. For example, consider the nature of debates that encircle the following questions: What are species? How did they come into being? How does natural selection shape patterns of variation in natural populations? Why did sex evolve? How did modern humans evolve? These are long-standing questions of broad interest, yet it could be difficult to make the case that they are congealed enough to merit a study of the ways we cannot answer them. For each example, one would have to begin an inquiry on the limits of knowledge by laying out a precise meaning of the question, and in each case there is little reason for confidence that others would agree with the way the question had been posed.

## SCIENTIFIC REVOLUTIONS AND HOW WE DO NOT SEE THEM

Regardless of one's view of the causes of scientific revolutions, they happen to all fields of inquiry. In the case of evolutionary biology, a good case can be made for at least two large turnovers since the time of Darwin's (1859) publication of *The Origin of Species*. The first is the flurry of questions and research that followed the rediscovery of Mendel's work, and the second is the modern synthesis (Provine, 1971). One of the most interesting features of these and other scientific revolutions is how investigators at the time could not anticipate them or even necessarily recognize when they were in the midst of them.

One of the best examples of scientific revolutions and of scientists' inability to anticipate them comes from physics; it hinges on the life and works of James Clerk Maxwell, the brilliant Scottish mathematician and physicist. Maxwell lived from 1813 to 1879 and was arguably the most capable of 19th-century physicists. Though he made many extraordinary contributions, his greatest legacies are the set of differential equations that describe the propagation of electromagnetic waves and the discovery that light and electromagnetic waves are both parts of the same thing. Maxwell's insight was generally far beyond his colleagues, and he was quite capable of simply sitting down and solving some well-posed problem of mathematical physics that had defeated everyone else. In 1871, he said: ". . . that, in a few years, all great physical constants will have been approximately estimated, and that the only occupation which will be left to men of science will be to carry these measurements to another place of decimals" (Harman, 1989, p. 244).

These words, however, were to be undermined by Maxwell's own discoveries. His findings about light and electromagnetism implied that electromagnetic waves propagated at a constant speed regardless of the speed of the observer. This created a conflict, eventually solved by Einstein with his theory of special relativity, that if the speed of light was constant, then velocities could not be additive, which they are under Newtonian mechanics.

Another player in the resolution of the conflict was Albert Michelson, who conducted the famous Michelson–Morley experiment, first in 1888, which helped to show that the speed of light did not depend on the speed of the observer. Curiously, Michelson made a statement during the dedication of the Ryerson Physics Laboratory at the University of Chicago in 1894 that was very similar to Maxwell's:

> The more important fundamental laws and facts of physical science have all been discovered, and these are now so firmly established that the possibility of their ever being supplanted in consequence of new discoveries is exceedingly remote. . . . Our future discoveries must be looked for in the sixth place of decimals. (Bernard, 1960, p. 123)

These brilliant scientists had no idea what was in store for their field, nor how their most basic concepts on natural laws were fundamentally wrong or at best sorely incomplete. In 1904, Einstein resolved the paradox that Maxwell had started, with the theory of special relativity. However, special relativity in turn generated a contradiction in Newton's theory of gravitation, which until then had been very successful. To resolve this, Einstein developed a new theory of gravity, the theory of general relativity. In this theory, gravity is not an instantaneously propagating force between masses, but rather a manifestation of the curvature of space–time, which in turn is shaped by the distribution of energy and momentum.

Newtonian mechanics and gravitation were wonderfully successful theories that did not quite work. Einstein's view of the universe worked much better—it even predicted the existence of black holes—but it was a very different universe than the one Maxwell and Michelson thought they lived in.

Maxwell's insight and lack of prescience extended even further. His theory of electromagnetism also revealed conflicts between electromagnetic theory and the theory of thermodynamics. The resolution of these conflicts, by Plank, Bohr, Heisenberg, Schrödinger, and de Broglie, took some time, but eventually led to what is now called quantum mechanics. During the problem solving, the view of the electron went from being literally a discrete subatomic particle to being literally a probability cloud.

These theories, special and general relativity and quantum mechanics, hold little resemblance to the physics of the 19th century that they replaced. The "knowledge" held by great scientists like Maxwell and Michelson within a few years was shown to be mostly wrong. The models that came to the fore and that persist were of an almost impossibly strange universe. Ironically, even as these scientists were laying the foundation for the revolution to come, they were boasting of the refined state of their science and elaborating on the limits of the knowledge.

If that kind of transformation can happen to as highly refined a science as physics of the late 19th century, then what transformations of knowledge might still lie in store for physics and other fields? In evolutionary genetics, what kind of revolution would be on a scale comparable to what occurred in physics? We could hope for any number of apparently impossible things. How about finding out that history also leaves an imprint, a better imprint, in some other media besides DNA or rocks. Or suppose time travel became possible for some totally unforeseen reason? Do these things seem any more impossible than the revolutions that occurred in physics? If for some reason we knew a revolution was coming, would we spend time on investigating the limits of knowledge under our current models?

By way of example, this physics history has served two points. The first is that scientific revolutions can happen regardless of our confidence in the anchors of what we presently call knowledge. The second is that we may not see the scientific revolutions coming. Fallible scientists sometimes miss the scientific revolution that is beginning to boil in their midst. Failing this insight, they may even expound on the advanced state of their science or elaborate on the limits of their knowledge.

## SCIENTIFIC REVOLUTIONS IN PROGRESS

Not all scientific revolutions are big, and much of the turnover of knowledge and questions that occurs in a field of inquiry does not deserve to be called a "revolution." Yet perhaps we can look at the state of our field and see signs of revolutions (small or big) in progress. The examples from physics suggest this may be difficult, but at the very least we can draw from these historical lessons and try to avoid a strong presumption of the security of our current knowledge. Similarly, there is nothing to prevent us from looking for rumbles of revolutions within our midst, inconsistencies and puzzles that may be the seeds of future turnovers.

Like any quickly moving field, evolutionary genetics is in flux with all manner of minirevolutions happening at any one time. However, there is at least one revolution in progress that seems to portend something larger on the horizon, and so it deserves some telling. This revolution hinges on the recent findings that natural selection plays the major role in shaping patterns of DNA sequence variation within and between taxa. There are two distinct components to these findings. The first are the many rejections of the neutral model that are based on commonly used neutral models and selective alternatives. The second component consists of observations, coupled with the reawakening of some old theory, that do not fit well into any of the standard neutral–selective paradigms (Hey, 1999).

## Rejecting the Neutral Model

In the first category are the numerous reports of rejection of neutral model predictions, all of which fall in the grand tradition of the neutralist–selectionist debate. Repeatedly, we see that the null model (neutrality and sometimes other assumptions) cannot explain a pattern in the data, thus lending statistical support to historical models that include natural selection. There are now dozens of reports of clear rejection of the neutral model. Some date to allozyme days, but the majority have come from applying new tests to comparative DNA sequence data. These observations in turn fall into two major categories. There are those tests that rely on both a specific population model (generally the Fisher–Wright population model) and the neutral infinite sites mutation model. Examples include the tests of Tajima and Fu and Li (Tajima, 1989; Fu and Li, 1993) as well as the famous Hudson Kreitman Aguadé (HKA) test (Hudson *et al.*, 1987). These tests have turned up a number of null model rejections, but they are sensitive to a number of things besides selection. Then there are those tests that do not depend on a specific population model. These include the McDonald–Kreitman test (McDonald and Kreitman, 1991) and a variety of ad hoc tests that have relied on other predicted covariates of selection (Begun and Aquadro, 1992; Kliman and Hey, 1993; Aquadro *et al.*, 1994; Akashi, 1994, 1996).

One of the most interesting and surprising aspects of these findings is that many of the conclusions regarding natural selection do not concern amino acid variation. It is now clear that natural selection plays a large role in determining codon usage in *Drosophila* (Akashi, 1994, 1995; Kliman and Hey, 1993, 1994; Akashi and Schaeffer, 1997). Also, since *Drosophila* introns are less variable and evolve more slowly than synonymous sites, it necessarily follows that intron sites also are under a fair bit of selective constraint.

## The Rediscovery of the Hill–Robertson Effect

The second component of what may be a scientific revolution in progress is the reawakening of some 30-year-old theory and the data that have inspired it. In 1966, Hill and Robertson studied the effect of linkage, between two sites each segregating two alleles under selection, on the probability of fixation of advantageous mutations. They found that under linkage, selection at one locus had a large impact on the probability of fixation at a second locus, and vice versa. In essence, the presence of linkage between two sites, each with alleles that varied in their contribution to fitness, caused the effectiveness of natural selection on both sites to be reduced. It is somewhat incongruous, but the more polymorphic sites that are added in linkage, the more poorly the process of natural selection acts to increase the frequency of the better alleles at each site. The effect is analogous to an acceleration in the rate of random drift: When more polymorphic loci are added in linkage, it is as if there were a reduction of the effective population size experienced by each locus (Hill and Robertson, 1966).

This work did not play a large role in the neutralist–selectionist debate, though it was certainly relevant (Lewontin, 1974), and until recently it has been absent from the molecular evolution literature. It was, however, Felsenstein's major reference in his seminal papers on the evolutionary advantage of sex (Felsenstein, 1974; Felsenstein and Yokoyama, 1976).

The basic idea is that selection at some sites adds an effectively random component to the variance in reproductive success that goes on at linked sites. In the days of population genetics, prior to the advent of large amounts of genotypic data, when allozymes and allele-based models ruled the day, the allozyme loci that were under study typically had large amounts of recombination between them. Rarely did authors find evidence of linkage disequilibrium between loci, and linkage effects were not a large component of our thinking with regard to natural selection. Today, with genotypic data and with more and more data sets emerging that describe lengthy haplotypes, the situation has changed and typically most of the polymorphisms reported in a study are tightly linked.

The observation that has pressed the Hill–Robertson issue is the one by Begun and Aquadro (1992), that polymorphism levels in *Drosophila melanogaster* correlate strongly with per generation recombination rates. This finding falls squarely in the domain of the Hill–Robertson effect: Genomic regions with less recombination experience more of the Hill–Robertson effect, as every polymorphic site is likely to be linked to many selected sites, and have a reduced effective population size and support reduced polymorphism levels. Today, the single most active area of research in *Drosophila*–theoretical population genetics is figuring out what kind of selection best explains this effect (Charlesworth, 1996).

The two main contenders are hitchhiking of selectively favored mutations and background selection against deleterious mutations. Both are models of strong selection and both are a limiting case of the more general Hill–Robertson effect. The hitchhiking model assumes neutrality for segregating variants and it envisions rare beneficial mutations that effectively run a broom through some portion of the genome, sweeping up all polymorphisms in a swath the width of which is defined by linkage (Maynard Smith and Haigh, 1974). The background selection model also assumes neutrality for segregating variants, at least those detectable in a sample of DNA sequences. In addition, a large proportion of all sequences are linked to deleterious mutations. Though each deleterious mutation is rare, there are many for large genomic regions of low crossing over, and so overall the effective population size can be greatly reduced to just that fraction that is not tightly linked to a deleterious mutation (Charlesworth *et al.*, 1993).

## The Seeds of a Revolution

What is sometimes overlooked in discussions of the relative merits of these models is that moderate or weak selection also can generate an appreciable Hill–Robertson effect. A mutation need not be strongly selected in order for selection to perceive it, especially if population sizes are large. Recall the growing heap of evidence that selection, acting within relatively small portions of the genome (e.g., the scope of a typical comparative DNA pop gene study), has played a large role in shaping variation. Recall also the very clear evidence that natural selection has a large impact on synonymous site variation, and that introns are not more variable and do not evolve faster than synonymous sites. In short, there is a large body of evidence that weak selection is acting on many of the polymorphisms that are segregating. It even seems possible that for large populations like *Drosophila*, selection may be able to detect essentially all the variation that is segregating.

The evidence for weak selection (e.g., on synonymous sites and intron sites) necessitates a closer look at the Hill–Robertson effect. In particular, it is worth asking, "How much of the observations that have been made and attributed to the Hill–Robertson effect may be due to weak selection?" The debate over Begun and Aquadro's (1992) observations have concerned the strong selection models, background selection, and selective sweeps (Charlesworth *et al.*, 1995; Hudson and Kaplan, 1995; Hamblin and Aquadro, 1996). Yet it may be that weaker selective effects have made a large contribution to the observation.

By way of example and to consider specifically the effects of weak selection under tight linkage, we can develop a model based upon the *Drosophila* dot chromosome. The euchromatic portion of this chromosome in *D. melanogaster* is probably about one megabase in length (Ajioka *et al.*, 1991). Most importantly, the chromosome experiences no crossing over, and polymorphism levels on it are sharply reduced relative to most regions of other chromosomes (Berry *et al.*, 1991). Let us ask the question, "What proportion of the observed drop in variation would be found if *all* mutations were weakly selected?" To examine this, assume that the rate of incoming mutations is the same as in the rest of the genome, and assume that all mutations are weakly selected, with a small value for the product of effective population size and selection coefficient ($s$). We can use estimates of 2 $Nu$ for the rest of the genome as a starting point for the population mutation rate experienced by the fourth chromosome ($N$ is the effective population size and $u$ is the mutation rate per generation). From a variety of genes, estimates for 4 $Nu$, tend to be around 0.005 per base pair for *D. melanogaster* (Moriyama and Powell, 1996). Assuming conservatively that this is twice the *total* (i.e., not just neutral) input of mutations every generation (i.e., $2 \times 2\ Nu$), then we can use the value of 0.0025 as the mutation rate per generation per base pair for the entire population. Then, for the entire megabase of the fourth chromosome we would expect there to be about $10^6 \times 0.0025 = 2500$ new mutations each generation.

A computer program was written that simulates a population of chromosomes that is receiving new mutations and from which samples are periodically assayed for polymorphism levels. Figure 1 shows results for the case of $2\ N = 50$, over a range of 2 $Ns$ values and 2 $Nu$ values. The highest value of 2 $Nu$ (2500) mimic the case described for the *Drosophila* fourth chromosome. Figure 2 shows the results for $2\ N = 500$. In both figures, it is clear that very weak selection ($2\ Ns \leq 1$) can have a large effect on polymorphism levels. The effect is slightly less for $2\ N = 500$ than for $2\ N = 50$, suggesting that $N$ is an important parameter in these considerations. Thus it is not clear from these results what proportion of the observed fourth chromosome effect (which arises from very large values of $N$ in natural populations) could be due to very weakly selected mutations. However, these results clearly show that weak selection and tight linkage can lead to a strong reduction in polymorphism levels. In practice, we have little insight on the strength of selection on synonymous sites and intron sites [but see Akashi and Schaeffer (1997)]. However, if they are under selection, they will contribute to a Hill–Robertson effect, and this effect may be considerable under high mutation rates and tight linkage. Also, all these simulations generated slightly to moderately negative values of Tajima's $D$ (Tajima, 1989), typically on the order of –0.5 (results not shown).

If indeed most segregating variants have some phenotypic effect and are under selection, and if they make an appreciable contribution to a Hill–Robertson effect, then there are several ideas held dear by evolutionary geneticists that may get overturned. There seem to be at least three noteworthy components of the scientific revolution in progress, and it remains to be seen how large they will grow:

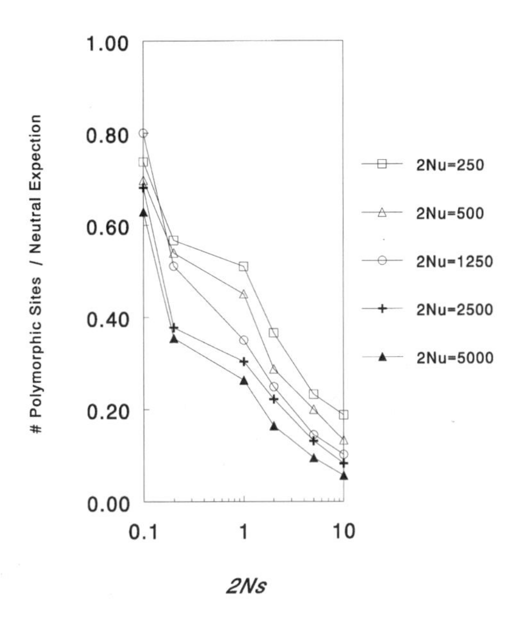


FIG. 1. The number of polymorphic sites, relative to that expected under neutrality, as a function of the strength of selection, 2 *Ns*, and the mutation rate, 2 *Nu*. Each point represents the mean of 100 measurements of the number of polymorphic sites in a random sample of 8 sequences. This mean value was divided by the expected number of polymorphic sites assuming neutrality of all mutations:

$$4\,Nu\sum_{i=1}^{n-1}\frac{1}{i}$$

for $n = 8$ (Watterson, 1975). Simulations were carried out using a constant population size of 50 chromosomes ($N = 25$), with no recombination. Mutations were added following an infinite sites model, each new mutation assigned to an unused segment of the genome (Kimura, 1969). Half of all mutations had a beneficial selection coefficient of *s* and half had a deleterious effect of *s*. Fitness was multiplicative across loci, with no dominance. In each generation following the addition of new mutations, individuals are grouped by fitnesses, and the numbers in each class in the next generation were generated by randomly sampling from a multinomial distribution having parameters that were the expected number of individuals in each fitness class following selection. The next generation was formed by randomly drawing (with replacement) the appropriate multinomial random number from each fitness class.

1. The ongoing debate over selective sweeps versus background selection may be too simplistic. At present, discerning the two models is very difficult, and the problem may become more difficult as we realize that other models also account for the data.
2. Our current search for estimates of the population neutral mutation parameter, 4 *Nu*, may be misguided. If all polymorphisms are under selection, then neutral model predictions are less interesting.

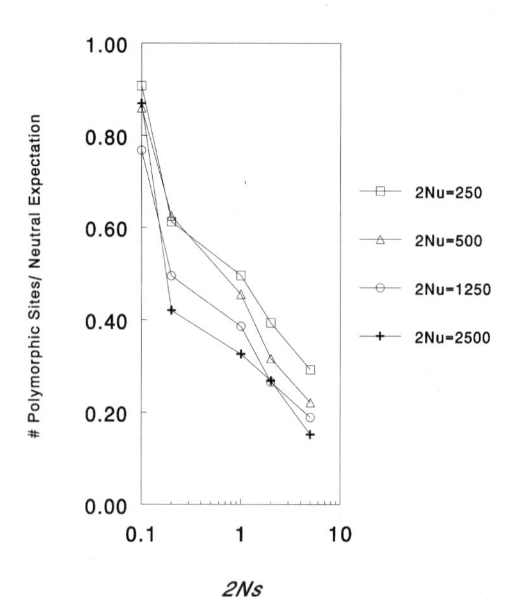


FIG. 2. Simulations results for a population size of 500 chromosomes ($N = 250$). See Fig. 1 legend for details.

Furthermore, if segregating polymorphisms are causing a Hill–Robertson effect, then they are reducing the effective population size for that portion of the genome in which they segregate. Thus even the highest current estimates of 4 *Nu* may be a good bit lower than would be found with only neutral mutations and no Hill–Robertson effect.

3. What do we mean by "natural selection" when we think of it acting on DNA sequence variation? If most polymorphisms are under selection, and the Hill–Robertson effect is pervasive, then the conventional distinction between genetic drift and natural selection becomes completely entangled with itself. Consider this circle: (1) if there are multiple sites under selection and linkage, then the Hill–Robertson effect comes into play and there is more genetic drift; but (2) the faster genetic drift goes, then the fewer sites are perceived by natural selection and the more weakly selected sites become effectively neutral. The idea seems constitutively antithetical, as if "natural selection equals genetic drift." Actually it is close to this, but there is an important distinction to be made. It is not natural selection that causes more drift, but rather segregating functional variants. In short, when there are more of these under linkage, then natural selection works more poorly, necessarily, as there is a concomitant acceleration in the rate of genetic drift.

This last point may be the most significant in the context of scientific revolutions. In the time since Kimura taught us the neutral model, evolutionary geneticists have had two poles of evolutionary forces between which lay virtually all their thinking about the way that evolution worked. Natural selection has been the deterministic force, whereas genetic drift is the random force. There is a possibility that this paradigm will have to give way, as we find that the two are interlinked in ways we are only poorly prepared to think about.

## CONCLUSIONS

While it may be appropriate for scientists in some mature fields to look inward and develop a research program on the limits of their science, I do not think evolutionary genetic research of this type is worthwhile. In this chapter I have outlined three main reasons for this view.

In the first place, the large questions of evolutionary genetics are not very well posed, and this is for the simple reason that there is considerable disagreement among scientists on the meanings of a large portion of our lexicon. Any attempt to focus on the limits in pursuing a specific question would have to occur within a relatively small community who understood with unanimity the meaning of that question. In general, questions of this type will be small and so the value of insights to the limits of knowledge may not be very interesting.

But suppose we were to consider just those questions that are not subject to very much uncertainty and that seem to be clear in the same way (and for the same reasons) to an entire community of scientists. Certainly we perceive limitations on our capacity to answer questions of this sort, so why not explore these limitations? In this situation scientists can and often do focus on the limits of knowledge. However, even these efforts may prove pointless. The reason, and this is the second reason against dwelling on the limits of knowledge, is that scientific revolutions do occur. They can arrive suddenly, without foreshadowing, and when they do, the old knowledge goes into the history books. Imagine being a researcher focused tightly on the limits of knowledge in your field, when suddenly your firmament is overturned. Your work may not even make the history books. Imagine instead, spending one's energy bringing on the revolution; how much more potential for discovery, and fun, lie in the path of a research program that is focused on resolving the large primary questions in a science.

The third and final point is that evolutionary genetics is currently undergoing scientific revolutions, though most are probably small. Ongoing revolutions can be difficult to recognize, as they may not be quick, and it may be difficult to foresee the future impact of day-to-day discoveries. But I think a good case can be made that we may be in the midst of a fairly large scientific revolution. There is a chance that our current ways of thinking about natural selection and our usual distinctions between genetic drift and natural selection may be about to undergo a fairly large change. At least, it does not seem to be a good time to dwell on the limits of our current knowledge when there is the chance to find out that much of what we think we know is wrong.

## REFERENCES


Ajioka, J. W., Smoller, D. A., Jones, R. W., Carulli, J. P., Vellek, A. E., Garza, D., Link, A. J., Duncan, I. W., and Hartl, D. L., 1991, *Drosophila* genome project: One-hit coverage in yeast artificial chromosomes, *Chromosoma* **100:**495–509.

Akashi, H., 1994, Synonymous codon usage in *Drosophila melanogaster*: Natural selection and translational accuracy, *Genetics* **136:**927–935.

Akashi, H., 1995, Inferring weak selection from patterns of polymorphism and divergence at "silent" sites in *Drosophila* DNA, *Genetics* **139:**1067–1076.

Akashi, H., 1996, Molecular evolution between *Drosophila melanogaster* and *D. simulans*: Reduced codon bias, faster rates of amino acid substitution, and larger proteins in *D. melanogaster*, *Genetics* **144:**1297–1307.

Akashi, H., and Schaeffer, S. W., 1997, Natural selection and the frequency distributions of "silent" DNA polymorphism in *Drosophila*, *Genetics* **146:**295–307.

Aquadro, C. F., Begun, D. J., and Kindahl, E. C., 1994, Selection, recombination, and DNA polymorphism in *Drosophila*, in: *Non-Neutral Evolution* (B. Golding, ed.), pp. 46–56, Chapman and Hall, London.

Begun, D. J., and Aquadro, C. F., 1992, Levels of naturally occurring DNA polymorphism correlate with recombination rates in *D. melanogaster*, *Nature* **356:**519–520.

Bernard, J., 1960, *Michelson and the Speed of Light*, Doubleday, Garden City, New York.

Berry, A. J., Ajioka, J. W., and Kreitman, M., 1991, Lack of polymorphism on the *Drosophila* fourth chromosome resulting from selection, *Genetics* **129:**1111–1117.

Charlesworth, B., 1996, Background selection and patterns of genetic diversity in *Drosophila melanogaster*, *Genet. Res. Camb.* **68:**131–149.

Charlesworth, B., Morgan, M. T., and Charlesworth, D., 1993, The effect of deleterious mutations on neutral molecular evolution, *Genetics* **134:**1289–1303.

Charlesworth, D., Charlesworth, B., and Morgan, M. T., 1995, The pattern of neutral molecular variation under the background selection model, *Genetics* **141:**1619–1632.

Darwin, C., 1859, *The Origin of Species*, John Murray, London.

Felsenstein, J., 1974, The evolutionary advantage of recombination, *Genetics* **78:**737–756.

Felsenstein, J., and Yokoyama, S., 1976, The evolutionary advantage of recombination. II. Individual selection for recombination, *Genetics* **83:**845–859.

Fu, Y. X., and Li, W. H., 1993, Statistical tests of neutrality of mutations, *Genetics* **133:**693–709.

Hamblin, M. T., and Aquadro, C. F., 1996, High nucleotide sequence variation in a region of low recombination in *Drosophila simulans* is consistent with the background selection model, *Mol. Biol. Evol.* **13:**1133–1140.

Harman, P. M. (ed.), 1989, *The Scientific Letters and Papers of James Clerk Maxwell*, Vol. 2, Cambridge University Press, Cambridge, England.

Hey, J., 1999, The neutralist, the fly, and the selectionist, *Trends Ecol. Evol.* **14:**35–38.

Hill, W. G., and Robertson, A., 1966, The effect of linkage on limits to artificial selection. *Genet. Res.* **8:**269–294.

Hudson, R. R., and Kaplan, N. L., 1995, Deleterious background selection with recombination, *Genetics* **141:**1605–1617.

Hudson, R. R., Kreitman, M., and Aguadé, M., 1987, A test of neutral molecular evolution based on nucleotide data, *Genetics* **116:**153–159.

Kimura, M., 1969, The number of heterozygous nucleotide sites maintained in a finite population due to steady flux of mutations, *Genetics* **61:**893–903.

Kliman, R. M., and Hey, J., 1993, Reduced natural selection associated with low recombination in *Drosophila melanogaster*, *Mol. Biol. Evol.* **10:**1239–1258.

Kliman, R. M., and Hey, J., 1994, The effects of mutations and natural selection on codon bias in the genes of *Drosophila*, *Genetics* **137:**1049–1056.

Lewontin, R. C., 1974, *The Genetic Basis of Evolutionary Change*, Columbia University Press, New York.

Maynard Smith, J., and Haigh, J., 1974, The hitch-hiking effect of a favourable gene, *Genet. Res. Camb.* **23:**23–35.

McDonald, J. H., and Kreitman, M., 1991, Adaptive protein evolution at the *Adh* locus in *Drosophila*, *Nature* **351:**652–654.

Moriyama, E. N., and Powell, J. R., 1996, Intraspecific nuclear DNA variation in *Drosophila*, *Mol. Biol. Evol.* **13:**261–277.

Provine, W. B., 1971, *The Origins of Theoretical Population Genetics*, University of Chicago Press, Chicago.

Tajima, F., 1989, Statistical method for testing the neutral mutation hypothesis by DNA polymorphism, *Genetics* **123:**585–595.

Watterson, G. A., 1975, On the number of segregating sites in genetical models without recombination, *Theor. Pop. Biol.* **7:**256–275.

# III

# Limits to Historical Inference and Prediction

The study of evolution is, in part, the study of biological history. The evolutionary geneticist has an advantage over the student of human cultural history in three major respects. First, evolutionists can manipulate their system and in effect rerun history as they try to understand the underlying mechanisms that govern change. The second advantage arises because the rules of genetic transmission are regular and susceptible to mathematical analysis. It is possible to make mathematically precise predictions about gene transmission under various sets of assumptions. The third advantage lies in the ability to sort out the temporal sequence of mutational change, which provides a crude means of dating temporal events and of ordering organismal relationships within biological groups. This record is always with us because it is part of our genetic endowment; we do not have to work from the fragments of records that have survived eons of time as do the students of human history, although ultimately the genetic record also erodes, owing to the relentless accumulation of mutations.

The mathematical biologist Gary Churchill (Chapter 6) explores the problem of inferring an ancestral character state on a evolutionary tree as a specific case of the reconstruction of history from extant information. Churchill employs a Bayesian framework to analyze several cases that include a molecular clock assumption or that dispense with a clock assumption and that feature star phylogenies versus treelike phylogenies. Churchill ends with the observation that what we conclude from such an analysis "depends on what we believed to begin with." That is not to say that we cannot illuminate history, but rather there is a kind of relativistic limitation that includes our subjective choice of model and prior distribution, and these color the final conclusions.

Ultimately, our ability to infer the past is absolutely dependent on a genetic sample drawn from extant organisms (with the minor exception of ancient DNA), and Aquadro (Chapter 7) reminds us that sampling strategy imposes a practical limitation on knowledge. We cannot know about events not recorded in the sample. Aquadro also reminds us that mutation is a very slow phenomenon and the window of time resolved is limited by the slow accumulation of mutational change. Of course, as Aquadro notes, this problem may be mitigated by increasing sample size or by selecting DNA regions that evolve rapidly (e.g., microsatellite sequences). We also can anticipate major advances in our ability to acquire data, and this should reduce limitations associated with sample size.

Aquadro goes on to ask an important question: Have new phenomena been uncovered through research in molecular population genetics? He concludes that a new phenomenon resulting from the complex relationship between linkage and selection has emerged from molecular-level investigations. Specifically, the level of genetic diversity appears to be positively correlated with local recombination rate, and this is true for many genes and several different species. One may argue that this phenomenon was foreshadowed in theoretical work by Hill and Robertson and others (e.g., see Chapter 5, this volume), but it is important to acknowledge that there are phenomena waiting to be discovered, and present knowledge is certain to expand in ways that we may not anticipate.

Li and colleagues (Chapter 8) present a powerful example of comparative genomics in their analysis of primate color vision. Primate color vision is determined by a multigene family that includes an autosomal blue pigment gene and X-linked genes for red and green opsins. There are many tools available to the student of comparative genomics, and these are well illustrated in the chapter. The comparative approach is shown to be very powerful, because it exploits the accumulation of mutational change over extended periods of evolutionary time, and we learn, for example, that trichromancy has arisen independently by several different mechanisms in primate evolution. One particularly interesting finding is that gene conversion involving different gene family members has been important in the evolution of color vision genes, and in one case this may have produced a novel adaptive haplotype. But as Li and colleagues note, gene conversion also erases evolutionary history and renders certain details of the past unknowable. The authors also point out that there are fundamental problems with the measurement of biophysical parameters in this system that may limit precise knowledge about the way specific amino acid changes alter spectral sensitivity maxima.

The converse of historical inference is future prediction. The principles of evolutionary genetics are frequently used for prediction when the goals are utilitarian. For example, it is common for plant and animal breeders to estimate narrow heritibilities to predict short-term selection response. In recent years, students of phenotypic evolution also have applied these methods to the longer-term analysis of morphological evolution (see

Chapter 12, this volume). An area where prediction is especially important is conservation biology, where the goal is to develop strategies that will minimize threats to species survival based on the application of genetic, ecological, and demographic principles. Nunney (Chapter 9) explores this theme by focusing on effective population size ($N_e$) as a predictor of population survival. There are many problems with the estimation of effective population size that create uncertainties for prediction. Because the future is unknown, the conservation biologist must use uncertain estimates of the $N_e$ parameter to develop broad-based strategies that are robust to future uncertainty and to parametric uncertainty. This dictates general rather than specific guidelines in the development of conservation strategies.

Weir (Chapter 10) considers the uses of population genetic models for the estimation of human relationships and for the estimation of linkage between disease and marker genes. He makes the strong point that many applications in this area of inference confuse single-point estimates with mean values. The reliability of inferences derived from this subtle error is unknown, although it may be possible to use computer simulation to gain some insight into the question of reliability. Weir also reminds us of the fact that all estimates of relationship are relative to some assumed base population value. This relativity may be a fundamental limitation to knowledge because the true base population is lost in history and genealogical information is beyond the reach of science.

6

# Inferring Ancestral Character States

GARY A. CHURCHILL

## INTRODUCTION

We wish to address the question of how much information is available in a set of observable individuals to help us reconstruct the ancestral state of a character. The question can be framed in variety of contexts, but we focus on the following scenario. An investigator is studying a group of organisms that are related by a phylogenetic tree with a known topology and known branch lengths (presumably based on a sufficient amount of prior data). A new character is observed, and the investigator wishes to infer the state of the character at one or more of the internal nodes of the tree. This scenario was motivated by a study of fungus-growing ants for which a molecular phylogeny of the ant species is available (Fig. 1) and the character of interest is the type of fungus grown (T. Schultz, personal communication).

Without making any assumptions about the nature of the process of character change, it will be essentially impossible to make any statements about the ancestral states. We will assume some rather simple models of character change and investigate the precision with which an ancestral state can be recovered. It will become apparent that the more realistic (hence complex) models of character change provide lower precision for making inference. Thus, our results may be viewed as a best-case scenario.

It is natural to pursue questions of this type in a Bayesian inferential framework (Bernardo and Smith, 1997). Probability distributions can be

GARY A. CHURCHILL • The Jackson Laboratory, Bar Harbor, Maine 04609.

*Evolutionary Biology, Volume 32*, edited by Michael T. Clegg *et al.*
Kluwer Academic / Plenum Publishers, New York, 2000.

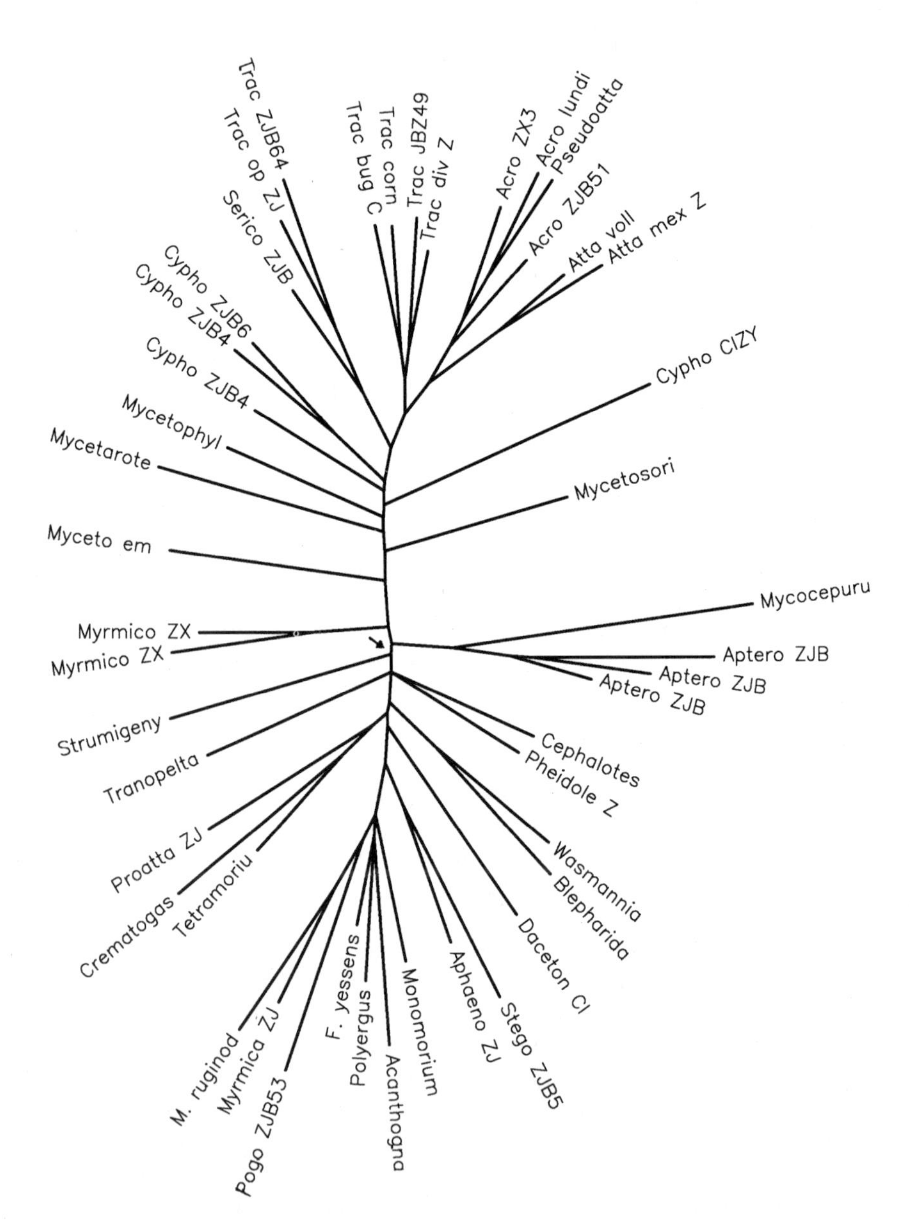


FIG. 1. Ant phylogeny, kindly provided by T. Schultz, showing the group of Attine ants (fungus growers) and other ant groups. The branch leading to the Attines is indicated by an arrow.

interpreted as descriptions of our state of knowledge (or uncertainty) regarding an unknown quantity. Bayes' rule provides a coherent means of updating our prior knowledge in light of new data. The usual arguments for classic likelihood methods do not apply in this data-limited setting. In some cases, standard arguments lead to nonsensible results and in general will lead one to overstate the confidence with which an ancestral character state can be inferred. Our goal here is to derive the posterior distribution of an ancestral character state, given observed data on extant individuals. The prior distribution in a Bayesian analysis is a matter of personal choice and often is the subject of criticism. For this reason we investigate the influence of the prior distribution on our inference.

# METHODS

## Model

We consider a generic character that can occur in one of two possible states, 0 or 1, neither having an *a priori* primitive or derived status. Our attention is restricted to binary characters with symmetric rates of exchange to simplify the analysis. We note that many morphological and behavioral characters are naturally binary (e.g., presence or absence). Extension of these results to multistate characters with general patterns of exchange may present some challenges, but the analytic framework remains essentially the same.

The character is assumed to evolve independently down each lineage of a known tree from an unknown ancestral state at the root of the tree. The probability that the character changes state at a given point in time depends only on its rate of change and on its current state. This is a standard Markovian model of evolution (Schultz *et al.*, 1996). The necessary concepts of continuous-time Markov processes can be found in Taylor and Karlin (1986) or any of a number of other basic texts. Our choice of a Markovian model is based on an assumption that the process of character state change is "memoryless." This assumption is likely to be violated in many situations. However, the alternative requires us to specify how the process "remembers" its past states. We proceed optimistically with the hope that our model is a sufficient approximation to reality so as not to be entirely misleading.

We consider star phylogenies in some detail. Due to the occurrences of radiations in the history of evolution, some phylogenies are nearly starlike. Of course, most phylogenies of interest are not starlike, and no set of individuals will ever have a truly starlike phylogeny. However, the star phylogeny provides a simple case on which to establish the principles of ancestral reconstruction. The star phylogeny results are compared to results obtained on a particular five-species tree topology. It appears that the star phylogeny can provide useful bounds on ancestral state probabilities and perhaps even an approximation for highly structured phylogenies, such as the tree in Fig. 1.

The observable outcome of the process of character evolution down a phylogeny is a character state configuration on $n$ observed individuals (e.g., {00111}). The observed configuration is the data from which we will infer ancestral character states.

It would be desirable to generalize this model. Even this very simple model provides significant insights into our ability to reconstruct ancestral character states. One question of interest is whether there is a star representation that is somehow equivalent to a given phylogeny. If this is the case, the star representation would provide a useful tool for partitioning the analysis of large phylogenies into manageable pieces.

## Prior Distributions

The prior distribution of a binary character state at the root node of a tree is

$$\begin{aligned} \Pr(s_0 = 0) &= \pi_0 \\ \Pr(s_0 = 1) &= 1 - \pi_0 \end{aligned} \tag{1}$$

The probability $\pi_0$ can be specified arbitrarily and is not necessarily the steady-state probability of the Markov process.

Defining a prior distribution on the rate of evolution is a more subtle and challenging problem. We need to specify a distribution Pr ($\theta$) that captures our notion of what reasonable rates of change might be. In this chapter we use a gamma distribution with density function

$$\Pr(\theta) \propto e^{-\lambda\theta}\theta^{\alpha-1}, \quad \theta > 0 \tag{2}$$

to represent our prior knowledge. The hyperparameters $\alpha$ (shape) and $\lambda$ (scale) are known quantities that are specified by the investigator. The mean of the gamma distribution is $\alpha/\lambda$. This can be interpreted as the *a priori* expected number of substitutions per unit time. The gamma distribution becomes more concentrated around its mean for large values of $\alpha$. Thus $\alpha$ is a measure of the strength of conviction in our prior knowledge. The gamma distribution is fairly flexible and yields analytically tractable results

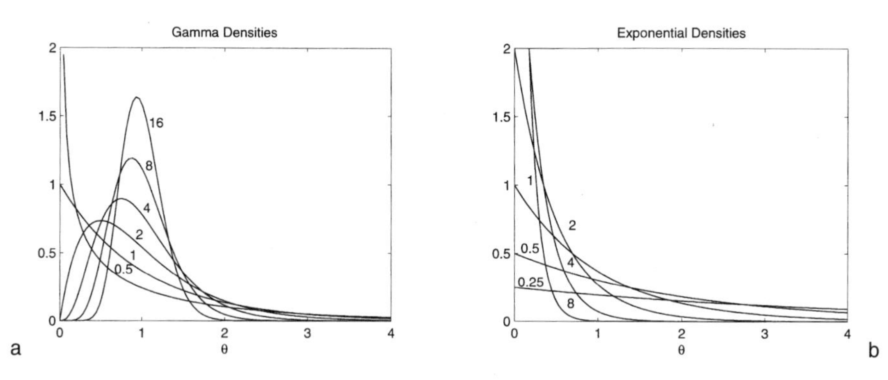


FIG. 2. Families of prior distributions are shown. Gamma densities with (a) mean one and (b) exponential densities.

for the problems considered here. Examples of gamma densities with mean one are shown in Fig. 2a. The exponential distribution arises as a special case of the gamma distribution when $\alpha = 1$. Examples of exponential densities are shown in Fig. 2b.

One goal of this investigation is to develop an understanding of how the choice of a prior distribution influences our confidence in the inference of an ancestral character state. The problem is naturally data limited in that it is not always possible to gather additional examples of extant individuals that are descendants of the common ancestor of interest. Furthermore, it may be difficult to extract much information regarding the rate of evolution of a character from the observable data. The extent to which observable data can dominate a prior belief represents a limit to our ability to draw conclusions regarding ancestral character states.

## Posterior Distributions

The application of Bayes' rule to derive the posterior distribution of an ancestral state is straightforward. The conditional distribution of $s_0$, given $s_1, \ldots, s_n$ and $\theta$, is

$$\Pr(s_0 \mid s_1, \ldots, s_n, \theta) \propto \Pr(s_0) \Pr(s_1, \ldots, s_n \mid s_0, \theta) \tag{3}$$

The proportionality constant is simply the sum over the possible values of $s_0$. The first term is the prior distribution of the ancestral state and the second term is conditional distribution of the data, given $\theta$ and the ancestral state.

The posterior distribution is a weighted average of the conditional distribution over values of $\theta$. Thus,

$$\Pr(s_0 \mid s_1, \ldots, s_n) \propto \Pr(s_0) \int \Pr(s_1, \ldots, s_n \mid s_0, \theta) \Pr(\theta) d\theta \tag{4}$$

The posterior distribution provides a complete description of our state of knowledge regarding the ancestral state $s_0$. It is a combination of information in the data with our prior knowledge regarding the ancestral state. Uncertainty due to the unknown rate of character change is accounted for by the weighted averaging. In the next section we derive the posterior distribution for various specific scenarios and examine the effects of our assumed prior knowledge on our conclusions.

## Computing

Numerical integrations were executed in MATLAB using a quadrature method. Most of the results were obtained analytically and confirmed by simulations. Source code is available from the author upon request.

# RESULTS

## One Branch

When the character states at two nodes connected by a branch in the tree are different, we say that a change of state has occurred across that branch. Results for the probability of a change of state across one branch form the basis for developments below. The probability of a change across a branch of length $t$ for a character evolving at rate $\theta$ is

$$\Pr(s_2 \mid s_1, \theta) = \begin{cases} \frac{1}{2}(1 + e^{-2\theta t}) & s_1 = s_2 \\ \frac{1}{2}(1 - e^{-2\theta t}) & s_1 \neq s_2 \end{cases} \tag{5}$$

This follows from properties of the Markov process. Over one unit of time ($t = 1$) the number of changes of state (events) that occur will have a Poisson distribution with mean $\theta$. If the number of events is odd, then $s_1 \neq s_2$, otherwise $s_1 = s_2$. Equation (5) is obtained by summing the odd (even) terms in the Poisson density. Note that multiple events are implicitly accounted for by the continuous-time Markov process.

Now suppose that $\theta$ is unknown but we have specified a prior distribution Pr ($\theta$). The unconditional probability of change is a weighted average over the prior distribution

$$\Pr(s_2 \mid s_1) = \int \Pr(s_2 \mid s_1, \theta)\Pr(\theta)d\theta \tag{6}$$

For the special case of a gamma prior, we obtain

$$\Pr(s_2 \mid s_1) = \begin{cases} \frac{1}{2} + \frac{1}{2}\left(\frac{\lambda}{\lambda + 2t}\right)^{\alpha} & s_1 = s_2 \\ \frac{1}{2} - \frac{1}{2}\left(\frac{\lambda}{\lambda + 2t}\right)^{\alpha} & s_1 \neq s_2 \end{cases} \tag{7}$$

Note that the time scale is arbitrary, and thus we can set $t = 1$.

This averaging process [Eq. (4)] is illustrated in Fig. 3. The conditional probability (given $\theta$) of a change of state over one unit of time is shown in Fig. 3a. The results of averaging these probabilities with respect to a family of prior distributions (exponential with mean $1/\lambda$) are shown in Fig. 3b. These latter probabilities reflect our uncertainty regarding the rate of change. The "flattening" of the curves in Fig. 3b reflects the fact that exponential prior distributions place a lot of weight on small rates of change. This assumption seems reasonable for many types of characters.

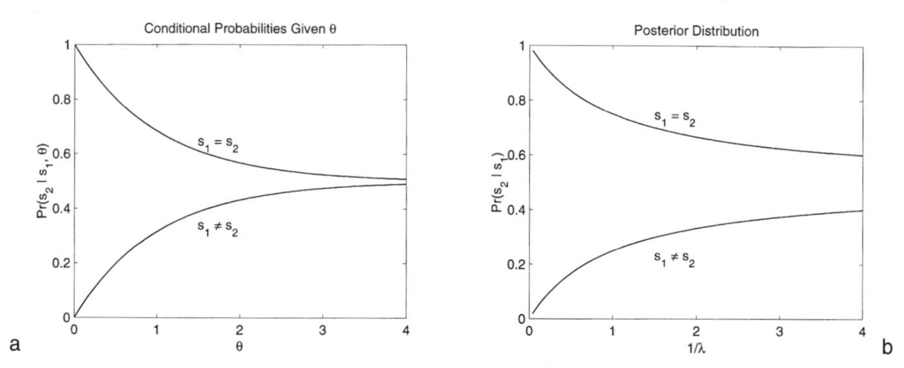


FIG. 3. Single branch transition probabilities $\Pr(s_2 \mid s_1, \theta)$ for the case of (a) known $\theta$ and (b) unconditional transition probabilities $\Pr(s_2 \mid s_1)$ averaged with respect to exponential prior distributions.

## Star Phylogeny with Clock

We consider a character that is observed on $n$ individuals related by a star phylogeny with a molecular clock (Fig. 4a). The clock assumption implies that all branches of the star have the same length $t = 1$. The conditional distribution of the root character state, given $\theta$ is

$$\Pr(s_0 \mid s_1, \ldots, s_n, \theta) \propto \Pr(s_0)\Pr(s_1 \mid s_0, \theta) \ldots \Pr(s_n \mid s_0, \theta)$$
$$\alpha \Pr(s_0)(1 - e^{-2\theta})^z (1 + e^{-\theta})^{n-z} \quad (8)$$

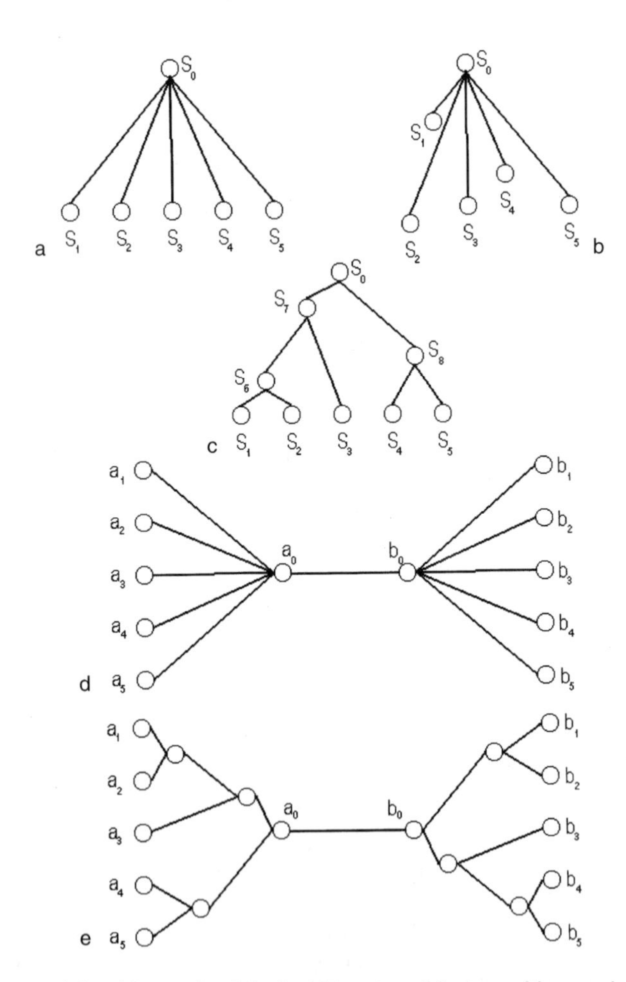


FIG. 4. Trees used in this study: (a) clocklike star, (b) star with no clock, (c) a five-species tree, (d) two stars, and (e) two trees.

where $z = \Sigma_{i=1}^{n} 1(s_i \neq s_0)$ is the total of number of changes from the ancestral state. Note that $z$ is a function of the ancestral state.

The posterior distribution of the root character state is

$$\Pr(s_0 \mid s_1, \ldots, s_n) \propto \Pr(s_0) \int_0^\infty \left(1 - e^{2\theta}\right)^z \left(1 + e^{-2\theta}\right)^{n-z} \Pr(\theta) d\theta \tag{9}$$

Despite its appearance, this integral is easy to solve numerically and for the gamma prior an analytic (if slightly messy) solution is available.

What does all this mean? Consider the case of the gamma prior distribution $\Pr(\theta) = \lambda e^{-\lambda\theta}\theta^{\alpha-1}$ for which the prior mean number of substitutions per unit time is $\alpha/\lambda$. Figure 5 shows the effects of the prior $\Pr(s_0)$ for sample sizes $n = 5$ and $n = 20$ and selected values of $\lambda$ and $\alpha$. The convergence of lines in the corners of these plots shows the effect of $\pi_0$ values near zero or one. For example, if you are certain *a priori* that the ancestral state was zero ($\pi_0 = 1$), then no amount of evidence in the data will sway you. Such an extreme view places us in the upper right corner of the graph. The data have a greater influence on our conclusions when the prior probability $\pi_0$ is more moderate (near 1/2). This is reflected in the maximal spread of the different curves. For the case $n = 5$, $\lambda = \alpha = 1$ (Fig. 5b) the configurations {00000} and {11111} provide strong evidence (values of the posterior probability are near zero or near one) regarding the ancestral state over a wide range of values of $\pi_0$. For intermediate configurations, such as {00111}, the evidence is weaker (values of posterior probability are nearer to 1/2). The effects of $\pi_0$ are modulated by our prior beliefs regarding the rate of change. When the rate change is believed to be high ($\lambda = 1/4$, Fig. 5c), our conclusions regarding the ancestral state are weakened. When the rate of change is believed to be low ($\lambda = 4$, Fig. 5a), our conclusions are strengthened to the extent that the off-by-one configurations (e.g., {01111}) now provide convincing evidence. Larger sample sizes with extreme proportions (e.g., 18 ones out of 20) can provide stronger evidence for the ancestral state (results not shown). The strength of our prior conviction (as reflected in the gamma shape parameter $\alpha$) can also influence our conclusions. In Fig. 5d, the prior mean rate of change is the same as in Fig. 5b. However, because the rate prior is more concentrated near the mean rate of one change per unit time, our conclusions regarding the ancestral state are weaker.

For high rates of change (e.g., $\lambda = 1/4$), even a small imbalance in the data provides a strong case for the ancestral state. When the rates are suspected to be lower (e.g., $\lambda = 4$), stronger evidence from the data is needed to draw conclusions about the ancestral state.

Figure 6 shows the effects of the rate prior on the posterior probability of the ancestral state for samples of size 5 and 20 and selected values of $\pi_0$ and $\alpha$. In the first two panels (6a and 6b), we see the case $\theta$ known. The

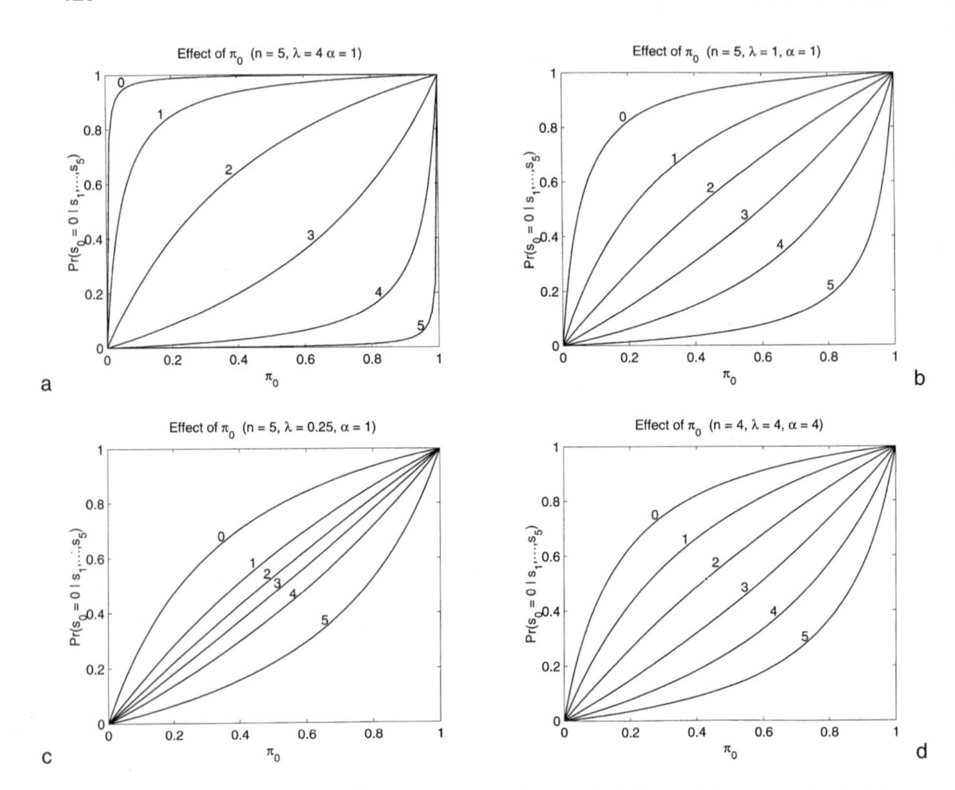


FIG. 5. Effect of $\pi_0$ = Pr ($s_0$) on the posterior probability of the ancestral state Pr($s_0$ | $s_1, \ldots, s_5$) for the clocklike star.

number of extant species with character state one is indicated next to each curve. The next two panels (6c and 6d) show the unconditional ($\theta$ unknown) posterior probabilities for the case of the exponential mean one prior. The "flattening" of these curves is analogous to that seen in Fig. 2. Ironically, not knowing $\theta$ actually strengthens our conclusions regarding the ancestral state for most of the data configurations shown here. In Fig. 6e we see the effect of shifting the state prior $\pi_0$. At higher rates of change our conclusions are drawn toward $\pi_0$. The effect of increasing $\alpha$ (Fig. 6f) is to bring the posterior probabilities closer to the case $\theta$ known. The effect of $\alpha$ is smaller when the sample size is larger (not shown).

## Star Phylogeny with No Clock

In this section, we consider the star phylogeny further but relax the molecular clock assumption (Fig. 4b). All extant species radiate

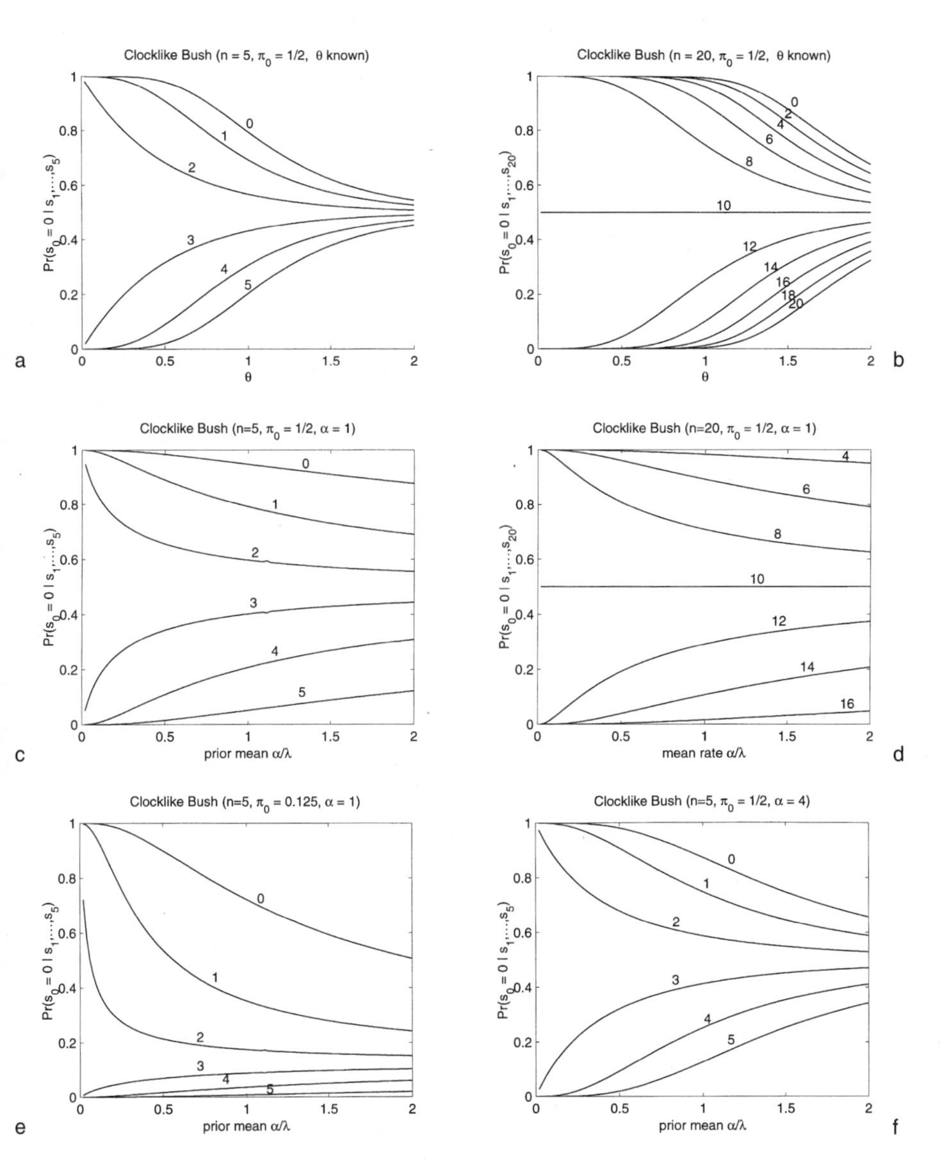


FIG. 6. Effect of the rate prior Pr(θ) on the posterior probability of the ancestral state $\Pr(s_0 \mid s_1, \ldots, s_5)$ for the clocklike star.

independently from a common ancestor, but the rate of evolution is allowed to vary between lineages. The lineage leading to individual $i$ has a rate $\theta_i$ for $i = 1, \ldots, n$. Rates are drawn independently from a gamma distribution.

The posterior distribution of the root character state is

$$\Pr(s_0 \mid s_1, \ldots, s_n) \propto \Pr(s_0) \prod_{i=1}^{n} \int \Pr(s_i \mid s_0, \theta_i) \Pr(\theta_i) d\theta_i$$

$$\propto \Pr(s_0) \left(1 - \left(\frac{\lambda}{\lambda+2}\right)^{\alpha}\right)^{z} \left(1 - \left(\frac{\lambda}{\lambda+2}\right)^{\alpha}\right)^{n-z} \quad (10)$$

Figure 7 shows the effects of varying $\pi_0$, $\lambda$, and $\alpha$ on the posterior probability of the ancestral state. These effects are similar to those for the bush model with the clock assumption. However, comparison of corresponding figures for models with and without the clock shows that conclusions regarding the ancestral state are less certain when the clock assumption is

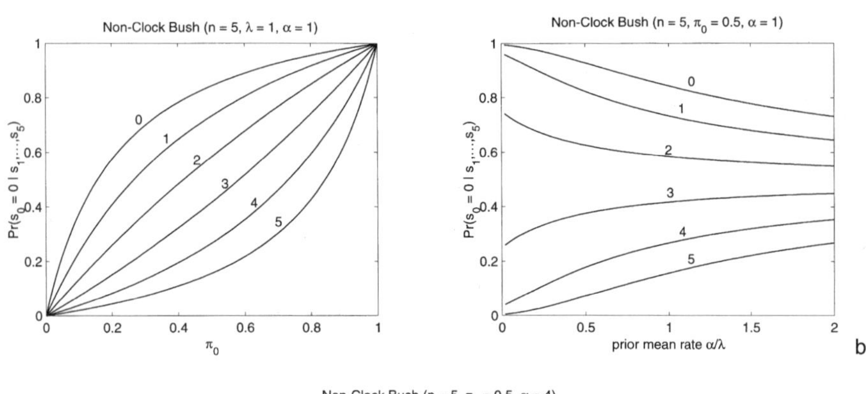


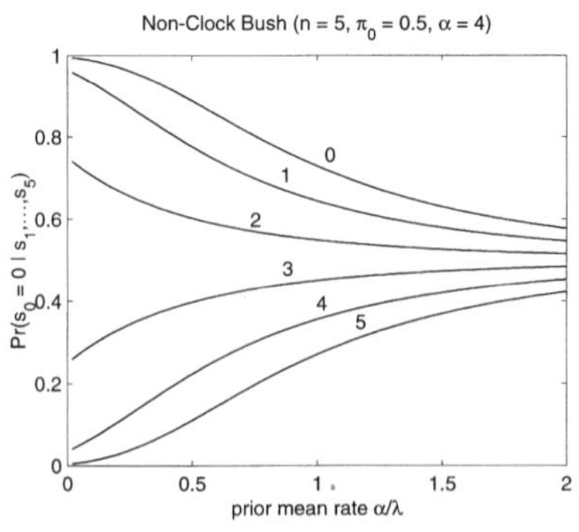


FIG. 7. Effect of the prior distributions $\Pr(s_0)$ and $\Pr(\theta)$ on the posterior probability of the ancestral state $\Pr(s_0 \mid s_1, \ldots, s_5)$ for the star with no clock.

relaxed. This reflects the fact that we cannot combine information about $\theta$ across the different lineages under this model.

## A CLOCKLIKE TREE

In this section we consider a tree phylogeny with five extant species (Fig. 4d). The lineages are no longer independent because some lineages share periods of common evolution. For a five-species tree, there are two internal nodes in addition to the root node. In general we may be interested in any or all of the ancestral states at these nodes (i.e., the joint distribution). With the clock assumption in effect, we can specify the root of the tree and will focus our developments on reconstructing the root character state. We assume that the branch lengths $t_1, \ldots, t_8$ are known and are proportional to time. The same (unknown) rate constant $\theta$ applies to all branches of the tree. The tree is scaled so that the total time from root to tip along any lineage is one.

Conditional probabilities (given $\theta$) for each of the $2^5 = 32$ possible configurations can be computed according to the results of Felsenstein (1981).

$$\begin{aligned}\Pr(s_1,\ldots,s_5 \mid \theta) = \sum_{s_0} \Pr(s_0) & \\ \times & \left( \sum_{s_8} \Pr(s_8 \mid s_0,\theta) \Pr(s_4 \mid s_8,\theta) \Pr(s_5 \mid s_8,\theta) \right) \\ \times & \left( \sum_{s_7} \Pr(s_7 \mid s_0,\theta) \Pr(s_3 \mid s_7,\theta) \right. \\ \times & \left. \left( \sum_{s_6} \Pr(s_6 \mid s_7,\theta) \Pr(s_1 \mid s_6,\theta) \Pr(s_2 \mid s_6,\theta) \right) \right) \end{aligned} \qquad (11)$$

Each sum extends over the two possible character states at an interior node. The single branch transition probabilities $\Pr(s_j \mid s_i)$ are given by Eq. (5). The posterior distribution is obtained by integrating this expression with respect to the prior distribution on $\theta$ as in Eq. (4).

There are 32 distinct tip configurations for the five-species tree. However, the clock assumption imposes constraints on the branch lengths such that there are only 18 distinct ancestral state probabilities. Table I shows the 32 configurations and the corresponding probabilities $\Pr(s_0 \mid s_1, \ldots, s_5)$. We are within 10% of certainty regarding the ancestral

TABLE I. Posterior Probabilities of Ancestral State $P = \Pr = (s_0 = 0|s_1, \ldots, s_5)$ for Each of the 32 Configurations on a Five-Species Tree (P tree) and the Five-Species Star Phylogeny (P star)[a]

| Configuration | P tree | P star |
|---|---|---|
| 00000 | 0.9101 | 0.9481 |
| 00100 | 0.6840 | 0.7922 |
| 01000 | 0.6796 | 0.7922 |
| 10000 | 0.6796 | 0.7922 |
| 00010 | 0.6762 | 0.7922 |
| 00001 | 0.6762 | 0.7922 |
| 11000 | 0.6513 | 0.5974 |
| 00011 | 0.5574 | 0.5974 |
| 01010 | 0.5429 | 0.5974 |
| 01001 | 0.5429 | 0.5974 |
| 10010 | 0.5429 | 0.5974 |
| 10001 | 0.5429 | 0.5974 |
| 10100 | 0.5284 | 0.5974 |
| 01100 | 0.5284 | 0.5974 |
| 00110 | 0.5128 | 0.5974 |
| 00101 | 0.5128 | 0.5974 |
| 11001 | 0.4872 | 0.4026 |
| 11010 | 0.4872 | 0.4026 |
| 10011 | 0.4716 | 0.4026 |
| 01011 | 0.4716 | 0.4026 |
| 10110 | 0.4571 | 0.4026 |
| 10101 | 0.4571 | 0.4026 |
| 01110 | 0.4571 | 0.4026 |
| 01101 | 0.4571 | 0.4026 |
| 11100 | 0.4426 | 0.4026 |
| 00111 | 0.3487 | 0.4026 |
| 11101 | 0.3238 | 0.2078 |
| 11110 | 0.3238 | 0.2078 |
| 01111 | 0.3204 | 0.2078 |
| 10111 | 0.3204 | 0.2078 |
| 11011 | 0.3160 | 0.2078 |
| 11111 | 0.0899 | 0.0519 |

[a] The prior distribution is set at $\pi_0 = 1/2$, $\lambda = 1$, and $\alpha = 1$.

state only for the fixed configurations {00000} and {11111}. The star phylogeny probabilities are shown in Table I for comparison. Notice that we come to almost within 5% of certainty under the star model.

The effects of the prior distribution on the posterior probability of the ancestral state are shown in Fig. 8. Figure 8a shows the effect of varying $\pi_0$. Notice that the 32 lines (18 are distinct) fall into six clusters corresponding to the numbers of ones at the tip of the tree. Reference lines for the

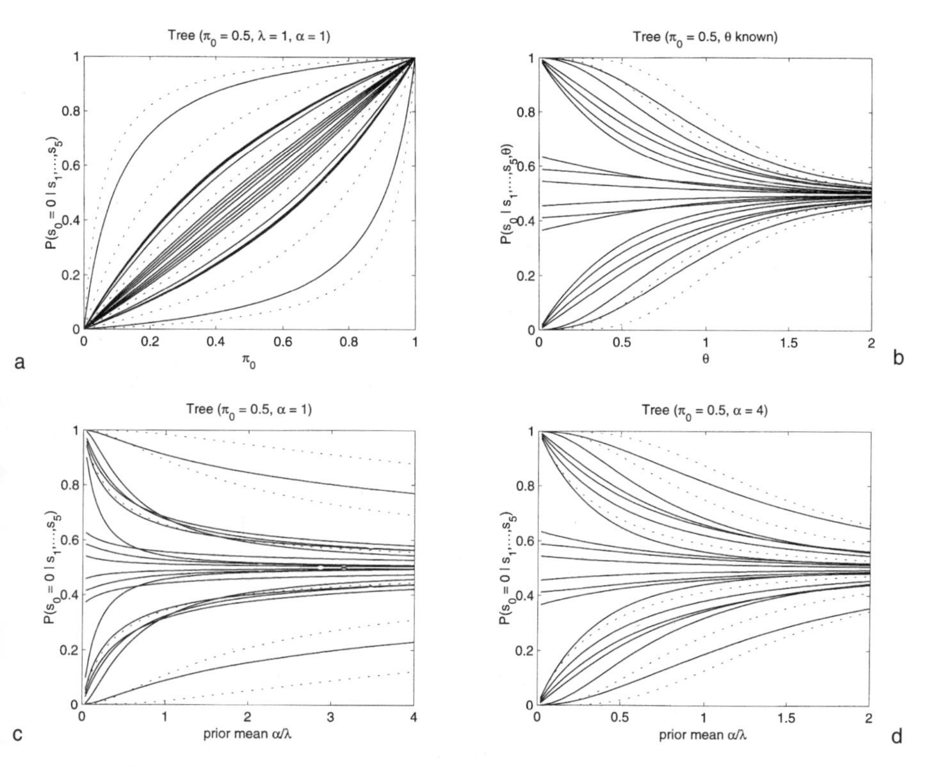


FIG. 8. Effect of the prior distributions $\Pr(s_0)$ and $\Pr(\theta)$ on the posterior probability of the ancestral state $\Pr(s_0 \mid s_1, \ldots, s_5)$ for the five-species tree.

clocklike bush ($n = 5$) show the loss of precision in estimation of the ancestral character state under the tree model. In Fig. 8b we see the case where $\theta$ is known. Figures 8c and 8d show the effects of varying the rate prior $\Pr(\theta)$. Again there are six clusters of lines, and the reference lines show the precision obtained under the bush model. Configurations {01001, 01010, 10001, 10010} and {01101, 01110, 10101, 10110} each require a minimum of two changes from the root state and are most strongly influenced by the prior. It is interesting that some of the lines cross as the prior mean rate is varied. This implies that the ordering of configurations with regard to their support of a given ancestral state changes as we vary the prior mean rate.

## Two Trees

In this section we consider a branch that is interior to a tree and compute the joint probability distribution of the character states at either end of this branch. Trees were constructed by connecting either two clock-like stars or trees at their roots by a central branch (Figs. 4d and 4e). We are interested in events that may have occurred on the central branch. This model may also be used to study the effects of adding an out-group to the set species under study.

Consider the four possible ancestral states $(a_0, b_0) = \{00, 01, 10, 11\}$ at both ends of the central branch. The observed data are $a_1, \ldots, a_m, b_1, \ldots, b_n$. The posterior distribution over ancestral states is given by

$$\Pr(a_0, b_0 \mid a_1, \ldots, a_m, b_1, \ldots, b_n) \propto \Pr(a_0)\Pr(b_0 \mid a_0) \times \Pr(a_1, \ldots, a_m \mid a_0)\Pr(b_1, \ldots, b_n \mid b_0) \tag{12}$$

where the first term is the prior probability of the ancestral state at the root of the $a$-tree, the second term is the single branch transition probability [Eq. (5)] and the last two terms are the bush or tree probabilities [Eq. (9) or Eq. (11) integrated as in Eq. (4)]. The probability of a change of state across the central branch is given by the sum of equation 12 over {01} and {10}.

Figures 9a and 9b show the effect of varying the rate prior on the central branch length for the two-bushes and two-trees models. The tip configuration of the $b$-tree is held constant at {11111}, and all possible con figurations are assigned to the $a$-tree. Probabilities for central branch configurations {00} and {01} for the most part are negligible. The probability of a change of state across the central branch increases with the prior mean rate $\alpha_c/\lambda_c$. Comparison of the two sets of figures reveals that the bush configuration gives greater precision of restoration as compared to a tree of the same size. Figures 9c and 9d show the effect of varying prior mean rate on the outer branches. It is not obvious from the figure that the curves are nonmonotone and converge to an asymptotic value of 1/2 for the probability of change at very high prior rates.

In Fig. 10 we examine the effect of varying the "out-group" size. Tree $a$ is a star of size 5, and tree $b$ is also a star, but the size is varied from zero (no out-group) to ten. All in-group species have character state zero and all out-group species have character state one. We see an increase in the probability of change on the central branch as we increase the prior mean rate $\alpha_c/\lambda_c$ (Fig. 10a). We see a decrease in the probability of change as the prior mean rate $\alpha_o/\lambda_o$ is increased (Fig. 10b). As the size of the out-group is increased, the probability of change increases. The increase is substantial for the first few additional members, but the returns diminish as the number increases much beyond five or six.

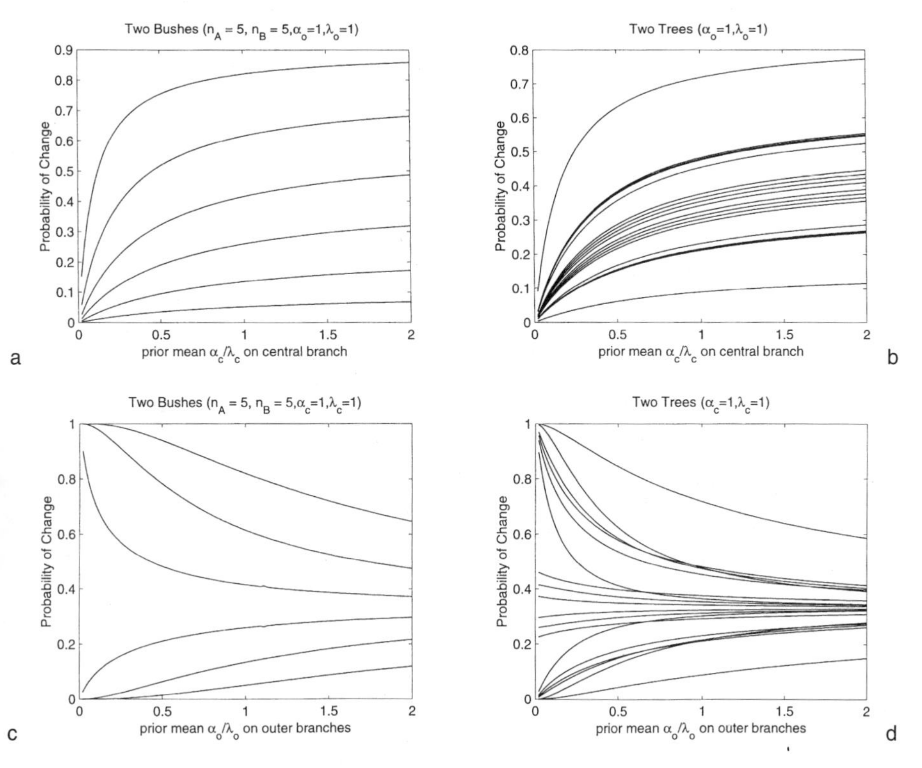


FIG. 9. Effects of varying the rate priors for the interior branch and exterior branches on the two-bushes (a, c) and two-tree (b, d) phylogenies.

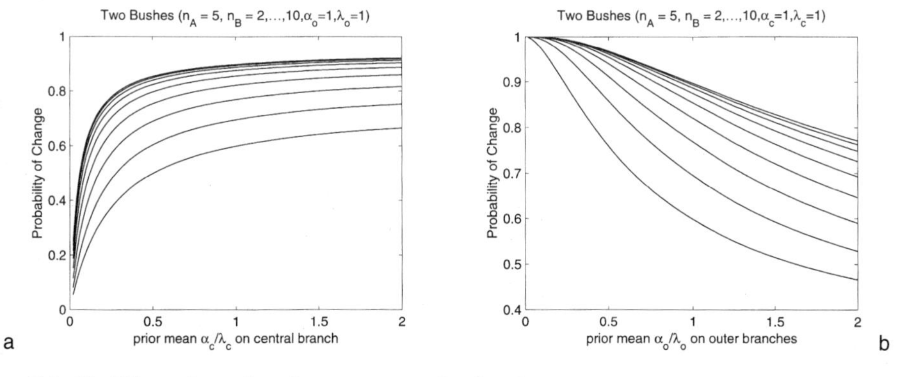


FIG. 10. Effect of varying the out-group size for the two-bushes (a) and two-tree (b) phylogenies.

## CONCLUSIONS

What are the sources of our uncertainty regarding an unknown quantity? Random variations in the data due to sampling fluctuations are straightforward to address, provided we have an accurate model of the process that generates the data. The classic theory of statistical inference (Lehmann, 1983) provides bounds (the Cramer–Rao lower bound or information inequality) on the precision with which an unknown quantity can be estimated.

More problematic perhaps is our uncertainty about the model itself. For example, convergent evolution due to correlated environmental conditions seems plausible for some morphological or behavioral traits. This scenario would violate the independent lineages assumption. However, without additional information it is difficult to know which model assumptions should be relaxed in any particular case. In light of this type of uncertainty, the results based on our Markov model provide a "best-case" analysis for assessing the reliability of an inferred ancestral state.

In summary, by investigating the effect of the prior distributions on ancestral state inference we have learned that in cases such as this one, where the amount of data is limited to a single character, what we conclude depends on what we believe to begin with. If our convictions are very strong, then no amount of evidence in the data will sway us. Whereas, for the "wishy-washy" among us, the slightest contradictions in the data can instill feelings of doubt. The situation is unavoidable, and the only honest course of action is to examine carefully the effects of prior assumptions and to make these clear when presenting conclusions.

## REFERENCES

Bernardo, J. M., Smith, A. F. M., 1997, *Bayesian Theory*, Wiley, New York.

Felsenstein, J., 1981, Evolutionary trees from DNA sequences: A maximum likelihood approach, *J. Mol. Evol.* **17:**368–376.

Lehmann, E. L., 1983, *Theory of Point Estimation*, Wiley, New York.

Schultz, T. R., Cocroft, R. B., and Churchill, G. A., 1996, The reconstruction of ancestral character states, *Evolution* **50:**504–511.

Taylor, H. M., and Karlin, S., 1986, *An Introduction to Stochastic Modelling*, Academic Press, London.

# 7

# The Problem of Inferring Selection and Evolutionary History from Molecular Data

CHARLES F. AQUADRO

## INTRODUCTION

Population genetics has gone through several periods of intense theoretical and empirical advancement, in part characterized by polarization of views, and each followed by what might be characterized as pessimism and frustration at the failure to arrive at definitive conclusions. For example, the discovery of extensive chromosomal polymorphism in *Drosophila* contributed to the polarized views in the 1940s and 1950s concerning the extent of genetic polymorphism in natural populations and to whether mutation pressure and drift or balancing selection maintained this variation (reviewed in Lewontin, 1974). Allozymes revealed a great wealth of genetic variation in populations, leading to debate as to whether selection could maintain all of this polymorphism. These data and the newly acquired amino acid sequence data motivated the development of the view that most evolution at the molecular level was driven by drift, with selection playing primarily a constraint role, eliminating strongly deleterious mutations, and only rarely playing a creative role in adaptive change (in relative terms compared to the total number of nucleotides in the genome) (Kimura, 1983). Largely because of the ease of allozyme data collection, the "static sample"

---

CHARLES F. AQUADRO • Department of Molecular Biology and Genetics, Cornell University, Ithaca, New York 14853.

*Evolutionary Biology, Volume 32*, edited by Michael T. Clegg *et al.*
Kluwer Academic / Plenum Publishers, New York, 2000.

approach (Lewontin, 1991) of population genetics grew at an enormous rate, leading to the documentation of levels of variation in a tremendous diversity of organisms. Alas, while allozymes gave valuable new insights into patterns of gene flow and population structure and differentiation, data on electromorph frequencies alone proved to be relatively unsuited to critically resolving the evolutionary forces responsible for their extensive variation, and the neutralist–selectionist debate lost much of its vigor (e.g., Ewens and Feldman, 1976; Gillespie, 1991).

The next wave of enthusiasm came with the advent of restriction mapping and then direct sequence analysis of DNA variation, the ultimate lower limit of genetic variation. Here, the potential density of variation, the ability to look at any part of the genome one wished, and the correlated evolutionary histories brought about by limited recombination between adjacent base pairs, promised new opportunities to resolve the relative contributions of various evolutionary forces to the patterns observed (reviewed in Kreitman and Akashi, 1995). Molecular data also have opened opportunities to dissect the genetic architecture of phenotype, both through trait mapping in pedigrees and through association mapping in populations (Lander and Kruglyak, 1995; Risch and Merikangas, 1996). These studies, together with functional studies of biological mechanism (e.g., Watt, 1994, Chapter 4, this volume; Powers and Schulte, 1998), have fueled new optimism that an understanding of the evolutionary process is within reach. Despite this enthusiasm, the problems have turned out to be difficult. For example, often selection has turned out to be subtle, and alternatives to strict neutrality become so open-ended as to be seemingly hopeless to test, particularly with static samples. The struggle to measure variation has been replaced by the struggle to measure selection. But the struggle to measure variation was technical. We now have the tools at hand to measure the most elemental aspect of heredity: the nucleotide sequence. Are the limits we face still technical, are they analytical, or are they conceptual? Are all of the "big" questions formulated with us now just struggling to find the answers? Are all of the questions answered that can be answered? Are we at the limits of our understanding of the population genetic mechanisms of the evolutionary process? Has all of the useful theory been discovered and explored? Are we at the limits of our ability to test our increasingly complex theory?

Clegg (Chapter 2, this volume) has stated that population genetics is now a mature science and argues that it is facing new limits to knowledge. He has outlined a classification of six reasons why answers to certain questions are "outside the bounds of scientific knowledge." Three reasons are that the knowledge is knowable in principle, but we currently are not able to obtain the answer for technical, experimental, or theoretical reasons. The other three reasons state that it is "unknowable" because of complexity or other confounding reasons. Clegg argues that our field has moved through at least much of the technical limitations and now is coming up more face to face with the "unknowable." Do I agree? In some ways, yes, but I remain much less convinced of how limiting the unknowables are, or that it should lead to pessimism in our field. I do not think this is the final frontier for population genetics. These issues, however, do emphasize the limitations of some approaches that have come to be the "bread and butter" approaches of our field. In particular, I think we are becoming increasingly aware of the limitations of the "static sample" research program. But there are other approaches to be combined with or used instead of this approach that hold great promise. I also think that we are still early enough in the application of molecular tools to natural populations that we are in for some major surprises. Parameters such as the mutation rate are increasingly coming into the grasp of experimentalists, as is detailed information on the distribution of recombination across genomes. Some parameters, however, remain in the abstract and "partially knowable," such as effective population size or even true census size. I agree with Clegg that we must be frank about the limitations. The challenge is this: How do we know whether answers are in fact unknowable?

In this chapter I discuss why evolution is so difficult to study and the ways in which molecular data have increased our insight into evolutionary history. I then discuss some of the recent surprises of population genetics that underscore the difficulty in saying what is unknowable. Next, I discuss the issue of signal to noise in tracing evolutionary lineages and what DNA sequencing and computer technology advances are likely to do for us in terms of making knowable what might seem unknowable. Finally, I speculate on whether the questions are wrong or whether it is our approach that is wrong. What is the future of our field?

## WHY IS EVOLUTION SO DIFFICULT TO STUDY EXPERIMENTALLY?

Unlike the study of development, for example, where the time scale is short in the experimentalists time frame, evolution of complex phenotypes is a process that in most multicellular organisms runs on a time scale that often far outlives the investigator. In addition, the starting and ending point in development is very narrow. For a particular organism, a fruit fly for example, we start every time with an egg and a sperm that fuse, and virtually every time we end up 12 days later with a completely recognizable adult fly. The same is not true for evolution. This is partly due to the fact that there is a stochastic element in evolution and partly due to the opportunist nature of evolutionary change and adaptation. Outcome is context dependent; dependent on the genetic makeup of the particular organisms involved and influenced by the biotic and abiotic setting in which the process is carried out. These factors traditionally have contributed to evolutionary biology not becoming a widely experimental science in terms of manipulative experiments. This is changing with the increased use of bacteria and viruses (e.g., Souza *et al.*, 1997; Bull *et al.*, 1997) where genetic change over thousands of generations can be readily monitored within months or a few years. However, these latter studies do not allow the study of diploid sexual dynamics that underlie evolution in humans, for example. For most organisms, population genetics has relied on inferences of what has happened in the past and the relative impact of various evolutionary forces, based on the distribution of variation in a static sample taken from one or more closely related populations at essentially a single point in time.

## RECAPTURING EVOLUTIONARY HISTORY

Theory developed in the last decade has provided insight into the structure of allele genealogies expected under various simple neutral models (equilibrium, expanding, and shrinking populations, as well as with subdivision) (reviewed in Hudson, 1990). The impact of diversity enhancing and directional selection also has been explored, and the simplest extremes of selection have been shown to have dramatic impacts on reshaping the genealogy, not only for the sites under selection but also for sites linked to the targets of selection. However, in our studies of natural populations our efforts to recover these genealogies (and thus infer the past evolutionary or selective history of a species or gene region) depends on there being sufficient mutational "signal" to distinguish the shape of the genealogy. We also are faced with the fact that in many cases very different evolutionary scenarios give overlapping expected distributions (despite having different expectations; this reflects the inherently stochastic nature of the evolutionary process). An example is inferences of population expansion from the distributions of nucleotide sequence or repeat sequence differences among alleles sampled from a population (e.g., Rogers and Harpending, 1992; Reich and Goldstein, 1998). Some hypotheses are more easily tested than others, because of the greater likelihood of sufficient signal. For example, in the case of population expansions, we expect all genes in a genome (including mitochondrial and nuclear genes) to show the signature of an expansion (an excess of rare variants and a single peak in the distribution of the number of nucleotide sequence differences among alleles). Thus, while one gene may be insufficient to distinguish among equilibrium and nonequilibrium, we can look at other unlinked genes, and the average trend should provide the necessary signal to detect the true demographic history.

## ARE THERE SIGNIFICANT NEW INSIGHTS TO BE GAINED IN POPULATION GENETICS?

Absolutely. Take, for example, the demonstration several years ago of a positive, genomewide correlation between DNA variation and regional rates of recombination in *Drosophila* (Begun and Aquadro, 1992). This and related observations (e.g., Aguadé *et al.*, 1989; Stephan and Langley, 1989; Begun and Aquadro, 1991; Berry *et al.*, 1991) and theory (particularly Kaplan *et al.*, 1989) opened up a whole host of experimental questions in population genetics and have contributed to an awakened interest in the genomic distribution of recombination. In addition, consideration of possible explanations for the correlation has revealed that the impact of removal of linked deleterious mutations from across the genome had been largely ignored in the population genetics of molecular variation (Charlesworth *et al.*, 1993). This has opened up a new rich area of inquiry that promises challenges to theoretical and experimental population genetics for some time.

Rigorously testing why the correlation exists, however, is not a simple task (e.g., Aquadro *et al.*, 1994; Aguadé and Langley, 1994; Stephan, 1994; Hudson and Kaplan, 1995; Charlesworth, 1996; Aquadro, 1997; Nachman, 1997; Nachman *et al.*, 1998). Has there been a recent selective sweep of a newly arising advantageous mutation (Kaplan *et al.*, 1989), or is variation low in a region because of background selection (Charlesworth *et al.*, 1993)? For any one gene, this is a very difficult task (and may in fact be unresolvable if adaptation has favored segregating polymorphism rather than a single new mutation) (e.g., Gillespie, 1994). However, I believe that we do have a chance at discerning, at least in broad strokes, the relative contributions of selection on adaptive versus deleterious variation in determining the correlation. Some types of tests rely on the frequency distribution of variation (e.g., Tajima, 1989; Fu and Li, 1993). Yet some types or time scales of positive selection yield frequency distributions of variants that are indistinguishable from others and from neutrality (which is also the expectation under background selection). For example, even with a catastrophic sweep, there is a relatively narrow window of time after the sweep where sufficient variation has accumulated but has not had time to drift to a neutral distribution, where Tajima's test will reject neutrality (Braverman *et al.*, 1995; Simonsen *et al.*, 1995). But there are other predictions that may help us resolve the relative contributions of adaptive versus background selection (and thus I feel that this difficult problem does fall in the knowable class). For example, the contrast of variation on X and autosomes has the potential to distinguish some forms of selection. Likewise, the analysis of microsatellite loci, with high mutation rates, in geographically different populations may help catch sweeps in the act (e.g., Schlötterer *et al.*, 1997; Schug *et al.*, 1998). The geographic distribution of nucleotide sequence variation in regions of low recombination also can allow discrimination of selective sweeps from background selection (e.g., Hamblin and Aquadro, 1997; Stephan *et al.*, 1998).

It recently has been argued that the deleterious mutation rate is lower than previously thought, and possibly too low for background selection to contribute significantly to the correlation between variation and recombination (Keightley, 1996; Fernández and López-Fanjul, 1996; García-Dorado, 1997; Keightley and Caballero, 1997; however, see Shabalina *et al.*, 1997). Thus the reasonably good fit seen to date (e.g., Charlesworth *et al.*, 1993; Hudson and Kaplan, 1995; Charlesworth, 1996; Hamblin and Aquadro, 1996; Wayne and Kreitman, 1996) between levels of variation and those predicted under background selection with a genomic deleterious mutation rate of 1 cannot be taken as rejecting adaptive models such as selective sweeps. Measurements of mutation rates and rates of recombination in natural populations of several species are thus important and increasingly possible tasks that will help us discriminate these hypotheses. Combining sequence polymorphism studies with functional studies of naturally occurring variants and differences fixed between species also should contribute significantly to our understanding of the nature and frequency of adaptive fixation.

## THE PROBLEM OF WEAK OR RECENT SELECTION

It has become apparent that in many cases selection may be quite weak (and complex). The impact on genealogies of these modes of selection can be quite limited. This makes distinguishing these selective genealogies even more difficult from neutral genealogies. It also has become evident that sometimes selection acts in such a way as to bring a new variant into a population to polymorphic frequencies, but not to fixation. Examples include superoxide dismutase (Hudson *et al.*, 1997) and 6-phosphogluconate dehydrogenase (Begun and Aquadro, 1994) in *D. melanogaster*. In addition, there are loci that we know have been the target of strong selection but for which our current statistical tests do not show a departure from neutrality. An example is the pattern of DNA sequence polymorphism at the resistance to Dieldrin (insecticide) locus (*Rdl*) in *D. melanogaster*, where the knowledge of gene function, the functional significance of the resistant variant, and the pattern of spatial sequence homogeneity of the resistant allele is testament to the impact of selection despite a lack of departure from neutrality with current statistical tests (V. Bauer, R. Roush, and C. Aquadro, unpublished observations). Complex and often weak selection will pose a serious challenge not easily surmounted by better theory or larger sample sizes alone. Although complementary functional studies significantly improve our understanding of the evolutionary process, for many genes the answers regarding the details of all evolutionary forces may be unknowable.

## HOW DO WE INCREASE THE SIGNAL?

Our ability to distinguish among different genealogies (and thus different evolutionary scenarios) is directly related to the number of mutations in a gene region (as reflected by the level of polymorphism segregating in a natural population). With base pair polymorphisms, some regions have very little variation; for example, on average two randomly chosen human chromosomes differ at only 6 in every 10,000 nucleotide sites, with some regions having nearly an order of magnitude *less* variation (Li and Sadler, 1991; Harding *et al.*, 1997; Hey, 1997; Clark *et al.*, 1998; Nachman *et al.*, 1998). The expected time to coalescence for a neutral genealogy for a population is 4 $N_e$ generations (Tajima, 1983). Thus, even if natural selection has acted in such a way as to lead to a balanced polymorphism in such a region, unless it is very old (greater than 4 $N_e$ generations), then we may not see any signal of the deeper genealogy associated with this diversity-enhancing selection. Likewise, selective sweeps associated with the fixation of a new mutation can result in a region with little or no variation present (a so-called "selective sweep"). In a region of the genome with moderate to high recombination, the window of the sweep can be quite narrow, which can point to the gene region on which natural selection acted. However, in regions of low recombination and for which levels of variation are already very low, there will be no discernible "footprint" of selection, since there is no variation to sweep away. The fixation of a single nucleotide substitution, however suggestive, provides no statistical power. Again, functional studies of the alternative variants may help in these cases but are not trivial to undertake.

One approach to increasing the signal has been to focus on regions of the genome that have a high rate of mutation. In particular, microsatellites have mutation rates on the order of $10^{-4}$ to $10^{-6}$ (compared to base pair substitutions, which occur at a rate of perhaps $10^{-8}$ per generation) (Schug *et al.*, 1998). This has proven useful in humans, where base pair substitutions are so few and far between (e.g., Bowcock *et al.*, 1994). It also has been fruitful to consider the haplotypes formed by multiple types of variants (including the rapidly mutating microsatellites) (e.g., Tishkoff *et al.*, 1996; Hammer *et al.*, 1998; Stephens *et al.*, 1998), yet even here certain loci will have patterns that are so complex that we will have no statistical power using available tests (or tests likely to be developed).

## ISSUES OF LINKAGE DISEQUILIBRIUM

Will we ever be able to say rigorously what shapes patterns of linkage disequilibrium in natural populations? The problem again is that many alternative evolutionary scenarios predict linkage disequilibrium, and the statistics used to measure linkage disequilibrium are inherently noisy. Again, the density of polymorphisms is limited in the genome of many organisms, thus limiting the statistical power to resolve alternative explanations (or to even be able to say with certainty what the level of linkage disequilibrium is). This is because linkage disequilibrium often appears very patchy in the genome, both on a large scale (10s of kilobases) and on a small scale (100s of base paris) (e.g., Aquadro *et al.*, 1992; Miyashita *et al.*, 1993; Schaeffer and Miller, 1993; Kirby and Stephan, 1996; Laan and Pääbo, 1997). A specific understanding of patterns of linkage disequilibrium may not be "unknowable" in general. It may be unknowable for some gene regions with low variability (variability is necessary to assess nonrandom associations). But the growing technical ability to obtain very large samples and to determine rates of recombination per unit physical length is likely to provide a new level of insight into the patterns of linkage disequilibrium seen across genomes and between species. Further understanding of secondary structure of messenger RNAs also may provide important insight into local patterns of linkage disequilibrium (e.g., Kirby and Stephan, 1996).

## WILL THE ABILITY TECHNICALLY TO OBTAIN LARGE SAMPLE SIZES EXTEND THE LIMITS OF WHAT WE CAN RIGOROUSLY DISTINGUISH?

Will the ability to get samples of hundreds or thousands of sequences allow us to resolve the fine details of how, when, and what kind of natural selection has acted at a locus? Will 500,000 single nucleotide polymorphisms (SNPs) allow us more rigorous mapping of phenotype to genotype in humans (Collins *et al.*, 1997)? Yes, in many cases, but for some questions, the answer will be no. Increased sample sizes will not substantially increase the precision of our estimates of variability, and the large number of nucleotide polymorphisms within large genes make statistical tests of association difficult (e.g., Clark *et al.*, 1998; Nickerson *et al.*, 1998). Further, as these authors point out, the density of SNPs proposed may be too low for many gene regions with high regional rates of recombination. However, for the estimation of frequency distributions, larger sample sizes are essential. The power of large sample sizes to distinguish alternative evolutionary hypotheses really has not been adequately investigated. In one instance where it has been studied (Simonsen *et al.*, 1995), a widely used test of an equilibrium-neutral model of molecular evolution required sample sizes of 50 or more for reasonable power. Virtually all samples to date have been on the order of 10 alleles. Thus, the conclusions about the weakness or absence of selection for many genes may be reversed once appropriately large sample sizes are obtained. It also is true that most currently available statistical tests of alternative hypotheses focus only on subsets of the data (that is, number of segregating sites, average pairwise differences, the number of substitutions on the tips of allele genealogies, etc). Consideration of more of the data at once is likely to provide more statistical power. It also is possible that larger amounts of data may reveal new "rules" (like the correlation between variation and recombination) that are currently unanticipated.

Tied in with larger samples is the ability to analyze these data. Several groups are developing methods that use all of the information contained in genealogies of sequences from populations as opposed to simple summary statistics like the number of segregating sites or nucleotide diversity (e.g., Griffiths and Tavaré, 1994; Kuhner *et al.*, 1995; Harding *et al.*, 1997). These simulation-based methods of analysis are up against computational limits at present, but these surely are to be solved by advances in computing. This will allow fuller exploration of the increased power these approaches may provide to detecting selection (at present, these analyses have been applied largely to neutral genealogies and inferences of coalescent time and the dating of neutral mutations).

## PROSPECTS FOR THE FUTURE

Are there limits to population genetics? Sure. For example, while static samples of DNA sequence variation can provide insight into historical selection and population processes, the power to reject alternative evolutionary hypotheses often is limited. Does that make future population genetic studies of static samples useless? No. There are limits to Mendelian crosses, but that does not limit the utility of doing crosses. We need to apply these methods in conjunction with new tools and approaches, including functional analyses, and some outside of the traditional limits of population genetics. Our job as biologists is not simply to ask interesting biological questions but to ask answerable questions. Are we asking the unanswerable? For some genes in some organisms, the answer may be yes. We are at the limit of resolution. We will never find any more signal (DNA sequence variation is the finest level of discrimination, so we cannot go further). However, we do not yet know how larger sample sizes or approaches that use more of the data may increase our ability to resolve alternative hypotheses.

Clearly, we must not be content to be preoccupied with static population samples alone. Functional analyses may help us understand potential fitness differences between variants but are more easily justified if there is some reason to believe that selection has acted in a gene region. However, we may not be able to disentangle the precise order of fixations or determine which of the differences were driven by selection at the locus in question. Nonetheless, the integration of population sampling with phylogenetic inference of ancestral states (e.g., Long and Langley, 1993; Messier and Stewart, 1997) is likely to expand the boundaries of the knowable in evolutionary biology, particularly when these studies can be combined with functional studies of ancestral states. And as for general principles, we may be able to go to other organisms for which there is more signal or for which the possibility of direct experimental tests over a few years period of time are practical. Bacterial chemostats and phage evolution are two examples in which the life span of the organism has made evolutionary hypotheses experimentally testable over time spans of thousands of generations. Some questions of short-term evolutionary processes can be (and have been) addressed in this way even in *Drosophila* and other complex organisms with short generation time. We also are gaining the technical tools to be able to measure empirically rates of recombination and mutation, and thus disentangle the notorious products $Nc$ (population size and rate of recombination) and $N\mu$ (population size and mutation rate) in population genetics. Tremendous challenges remain in how to connect the microevolutionary process with macroevolution and development where we are trying to understand the evolution of phenotypes for which multiple loci have contributed in often complex ways.

Returning to Clegg's (Chapter 2, this volume) original classification of limits to knowledge in evolutionary biology, I agree that many (but not all) of the technical impediments to knowledge have been or are being overcome. The complexity of the evolutionary process certainly limits our ability to tease apart many important evolutionary processes. But I remain optimistic that we have much to learn as we increase our data collection and computational abilities and incorporate a larger group of methodologies into the population geneticist's toolbox. Population genetics is not an end in itself but rather one of the perspectives to be brought to bear on biological questions. Our task is to be sure we know what questions we really want to ask. Do we really want to know what is possible in evolution or exactly how a particular morphology, behavior, or distribution of genetic variation came to be? The boundaries of knowable and unknowable are likely to be very different for these two perspectives.

### Acknowledgments

I thank Mike Clegg and the Sloan Foundation for the opportunity to meet and discuss the topic of limits to knowledge, as well as members of my laboratory and Ward Watt for thoughtful criticism of the ideas discussed in this chapter.

## REFERENCES

Aguadé, M., and Langley, C. H., 1994, Polymorphism and divergence in regions of low recombination in *Drosophila*, in: *Non-neutral Evolution: Theories and Molecular Data* (B. Golding, ed.), pp. 67–76, Chapman and Hall, London.

Aguadé, M., Miyashita, N., and Langley, C. H., 1989, Reduced variation in the yellow-achaete scute region in natural populations of *Drosophila melanogaster*, *Genetics* **122:**607–616.

Aquadro, C. F., 1997, Insights into the evolutionary process from patterns of DNA sequence variability, *Curr. Opin. Genet. Dev.* **7:**835–840.

Aquadro, C. F., Jennings, Jr., R. M., Bland, M. M., Laurie, C. C., and Langley, C. H., 1992, Patterns of naturally occurring restriction map variation, dopa decarboxylase activity variation and linkage disequilibrium in the *Ddc* gene region of *Drosophila melanogaster*, *Genetics* **132:**443–452.

Aquadro, C. F., Begun, D. J., and Kindahl, E. C., 1994, Selection, recombination, and DNA polymorphism in *Drosophila,* in: *Non-neutral Evolution: Theories and Molecular Data* (B. Golding, ed.), pp. 46–55, Chapman and Hall, London.

Begun, D. J., and Aquadro, C. F., 1991, Molecular population genetics of the distal portion of the X chromosome in *Drosophila*: Evidence for genetic hitchhiking of the *yellow-achaete* region, *Genetics* **129:**1147–1158.

Begun, D. J., and Aquadro, C. F., 1992, Levels of naturally occurring DNA polymorphism correlate with recombination rates in *D. melanogaster*, *Nature* **356:**519–520.

Begun, D. J., and Aquadro, C. F., 1994, Evolutionary inferences from DNA variation at the 6-phosphogluconate dehydrogenase locus in natural populations of *Drosophila*: Selection and geographic differentiation, *Genetics* **136:**155–171.

Berry, A. J., Ajioka, J. W., and Kreitman, M., 1991, Lack of polymorphism on the *Drosophila* fourth chromosome resulting from selection, *Genetics* **129:**1111–1117.

Bowcock, A. M., Ruiz Linares, A., Tomfohrde, J., Minch, E., Kidd, J. R., and Cavalli-Sforza, L. L., 1994, High resolution of human evolutionary trees with polymorphic microsatellites, *Nature* **368:**455–457.

Braverman, J. M., Hudson, R. R., Kaplan, N. L., Langley, C. H., and Stephan, W., 1995, The hitchhiking effect on the site frequency spectrum of DNA polymorphisms, *Genetics* **140:**783–796.

Bull, J. J., Badgett, M. R., Wichman, H. A., Huelsenbeck, J. P., Hillis, D. M., Gulati, A., Ho, C., and Molineux, I. J., 1997, Exceptional convergent evolution in a virus, *Genetics* **147:**1497–1507.

Charlesworth, B., 1996, Background selection and patterns of genetic diversity in *Drosophila melanogaster, Genet. Res.* **68:**131–149.

Charlesworth, B., Morgan, M. T., and Charlesworth, D., 1993, The effect of deleterious mutations on neutral molecular variation, *Genetics* **134:**1289–1303.

Clark, A. G., Weiss, K. M., Nickerson, D. A., Taylor, S. L., Buchanan, A., Stengard, J., Salomaa, V., Vartiainen, E., Perola, M., Boerwinkle, E., and Sing, C. F., 1998, Haplotype structure and population genetic inferences from nucleotide sequence variation in human lipoprotein lipase, *Am. J. Hum. Genet.* **63:**595–612.

Collins, F. S., Guyer, M. S., and Chakravarti, A., 1997, Variations on a theme: Cataloging human DNA sequence variation, *Science* **278:**1580–1581.

Ewens, W. J., and Feldman, M. W., 1976, The theoretical assessment of selective neutrality, in: *Population Genetics and Ecology* (S. Karlin and E. Nevo, eds.), pp. 303–337, Academic Press, New York.

Fernández, J., and López-Fanjul, C., 1996, Spontaneous mutational variances and covariances for fitness-related traits in *Drosophila melanogaster*, *Genetics* **143:**829–837.

Fu, Y.-X., and Li, W.-H., 1993, Statistical tests of neutrality of mutations, *Genetics* **133:**693–709.

García-Dorado, A., 1997, The rate and effects distribution of viability mutation in *Drosophila*: Minimum distance estimation, *Evolution* **51:**1130–1139.

Gillespie, J. H., 1991, *The Causes of Molecular Evolution*, Oxford University Press, New York.

Gillespie, J. H., 1994, Alternatives to the neutral theory, in: *Non-neutral Evolution: Theories and Molecular Data* (B. Golding, ed.), pp. 1–17, Chapman and Hall, London.

Griffiths, R. C., and Tavaré, S., 1994, Ancestral inference in population genetics, *Stat. Sci.* **9:**307–319.

Hamblin, M. T., and Aquadro, C. F., 1996, High nucleotide sequence variation in a region of low recombination in *Drosophila simulans* is consistent with the background selection model, *Mol. Biol. Evol.* **13:**1133–1140.
Hamblin, M. T., and Aquadro, C. F., 1997, Contrasting patterns of nucleotide sequence variation at the glucose dehydrogenase (*Gld*) locus in different populations of *Drosophila melanogaster*, *Genetics* **145:**1053–1062.
Hammer, M. F., Karafet, T., Rasanayagam, A., Wood, E. T., Altheide, T. K., Jenkins, T., Griffiths, R. C., Templeton, A. R., and Zegura, S. L., 1998, Out of Africa and back again: nested cladistic analysis of human Y chromosome variation, *Mol. Biol. Evol.* **15:**427–441.
Harding, R. M., Fullerton, S. M., Griffiths, R. C., Bond, J., Cox, M. J., Schneider, J. A., Moulin, D. S., and Clegg, J. B., 1997, Archaic African and Asian lineages in the genetic ancestry of modern humans, *Am. J. Hum. Genet.* **60:**772–789.
Hey, J., 1997, Mitochondrial and nuclear genes present conflicting portraits of human origins, *Mol. Biol. Evol.* **14:**166–172.
Hudson, R. R., 1990, Gene genealogies and the coalescent process, *Oxf. Surv. Evol. Biol.* **7:**1–44.
Hudson, R. R., and Kaplan, N. L., 1995, Deleterious background selection with recombination, *Genetics* **141:**1605–1617.
Hudson, R. R., Sáez, A. G., and Ayala, F. J., 1997, DNA variation at the *Sod* of *Drosophila melanogaster*: An unfolding story of natural selection, *Genetics* **136:**1329–1340.
Kaplan, N. L., Hudson, R. R., and Langley, C. H., 1989, The "hitchhiking effect" revisited, *Genetics* **123:**887–899.
Keightley, P. D., 1996, Nature of deleterious mutation load in *Drosophila*, *Genetics* **144:**1993–1999.
Keightley, P. D., and Caballero, A., 1997, Genomic mutation rates for lifetime reproductive output with lifespan in *Caenorhabditis elegans*, *Proc. Natl. Acad. Sci. USA* **94:**3823–3827.
Kimura, M., 1983, *The Neutral Theory of Molecular Evolution*, Cambridge University Press, Cambridge, England.
Kirby, D. A., and Stephan, W., 1996, Multi-locus selection and the structure of variation at the white gene of *Drosophila melanogaster*, *Genetics* **144:**635–645.
Kreitman, M., and Akashi, H., 1995, Molecular evidence for natural selection, *Annu. Rev. Ecol. Syst.* **26:**403–422.
Kuhner, M. K., Yamato, J., and Felsenstein, J., 1995, Estimating effective population size and mutation rate from sequence data using Metropolis-Hastings sampling, *Genetics* **140:**1421–1430.
Laan, M., and Pääbo, S., 1997, Demographic history and linkage disequilibrium in human populations, *Nat. Genet.* **17:**435–438.
Lander, E., and Kruglyak, L., 1995, Genetic dissection of complex traits: guidelines for interpreting and reporting linkage results, *Nat. Genet.* **11:**241–247.
Lewontin, R. C., 1974, *Genetic Basis of Evolutionary Change*, Columbia University Press, New York.
Lewontin, R. C., 1991, Electrophoresis in the development of evolutionary genetics: Milestone or millstone? *Genetics* **128:**657–662.
Li, W.-H., and Sadler, L. A., 1991, Low nucleotide diversity in man, *Genetics* **129:**513–523.
Long, M., and Langley, C. H., 1993, Natural selection and the origin of *jingwei*, a chimeric processed functional gene in *Drosophila*, *Science* **260:**91–95.
Messier, W., and Stewart, C. B., 1997, Episodic adaptive evolution of primate lysozymes, *Nature* **385:**151–154.
Miyashita, N. T., Aguadé, M., and Langley, C. H., 1993, Linkage disequilibrium in the *white* locus region of *Drosophila melanogaster*, *Genet. Res.* **62:**101–109.

Nachman, M. W., 1997, Patterns of DNA variability at X-linked loci in *Mus domesticus*, *Genetics* **147:**1303–1316.

Nachman, M. W., Bauer, V. L., Crowell, S. L., and Aquadro, C. F., 1998, DNA variability and recombination rates at X-linked loci in humans, *Genetics* **150:**1133–1141.

Nickerson, D. A., Taylor, S. L., Weiss, K. M., Clark, A. G., Hutchinson, T. G., Stengard, J., Salomaa, V., Vartiainen, E., Boerwinkle, E., and Sing, C. F., 1998, DNA sequence diversity in a 9.7 kb region of the human lipoprotein lipase gene, *Nat. Genet.* **19:**233–240.

Powers, D. A., and Schulte, P. M., 1998, Evolutionary adaptations of gene structure and expression in natural populations in relation to a changing environment: A multidisciplinary approach to address a million-year saga of a small fish, *J. Exp. Zool.* **282:**71–94.

Reich, D. E., and Goldstein, D. B., 1998, Genetic evidence for a Paleolithic human population expansion in Africa, *Proc. Natl. Acad. Sci. USA* **95:**8119–8123.

Risch, N., and Merikangas, K., 1996, The future of genetic studies of complex human diseases, *Science* **273:**1516–1517.

Rogers, A. R., and Harpending, H. C., 1992, Population growth makes waves in the distribution of pairwise differences, *Mol. Biol. Evol.* **9:**552–569.

Schaeffer, S. W., and Miller, E. L., 1993, Estimates of linkage disequilibrium and the recombinational parameter determined from segregating nucleotide sites in the alcohol dehydrogenase region of *Drosophila pseudoobscura*, *Genetics* **135:**541–552.

Schlötterer, C., Vogl, C., and Tautz, D., 1997, Polymorphism and locus-specific effects on polymorphism at microsatellite loci in natural *Drosophila melanogaster* populations, *Genetics* **146:**309–320.

Schug, M. D., Hutter, C., Noor, M. A. F., and Aquadro, C. F., 1998, Mutation and evolution of microsatellites in *Drosophila melanogaster*, *Genetica* **102/103:**359–367.

Shabalina, S. A., Yampolsky, L. Y., and Kondrashov, A. S., 1997, Rapid decline of fitness in panmictic populations of *Drosophila melanogaster* maintained under relaxed natural selection, *Proc. Natl. Acad. Sci. USA* **94:**13034–13039.

Simonsen, K. L., Churchill, G. A., and Aquadro, C. F., 1995, Properties of statistical tests of neutrality for DNA polymorphism data, *Genetics* **141:**413–429.

Souza, V., Turner, P. E., and Lenski, R. E., 1997, Long-term experimental evolution in *Escherichia coli*, V. Effects of recombination with immigrant genotypes on the rate of bacterial evolution, *J. Evol. Biol.* **10:**743–769.

Stephan, W., 1994, Effects of genetic recombination and population subdivision on nucleotide sequence variation in *Drosophila ananassae*, in: *Non-neutral Evolution: Theories and Molecular Data* (B. Golding, ed.), pp. 57–66, Chapman and Hall, London.

Stephan, W., and Langley, C. H., 1989, Molecular genetic variation in the centromeric region of the X chromosome in three *Drosophila ananassae* populations. I. Contrasts between the *vermilion* and *forked* loci, *Genetics* **121:**89–100.

Stephan, W., Xin, L, Kirby, D. A., and Braverman, J. M. 1998. A test of the background selection hypothesis based on nucleotide data from *Drosophila ananassae, Proc. Natl. Acad. Sci. USA* **95:**5649–5654.

Stephens, J. C., Reich, D. E., Goldstein, D. B., Shin, H. D., Smith, M. W., Carrington, M., Winkler, C., Huttley, G. A., Allikmets, R., Schriml, L., Gerrard, B., Malasky, M., Ramos, M. D., Morlot, S., Tzetis, M., Oddoux, C., Di-Giovine, F. S., Nasioulas, G., Chandler, D., Aseev, M., Hanson, M., Kalaydjieva, L., Glavac, D., Gasparini, P., Kanavakis, E., Claustres, M., Kambouris, M., Ostrer, H., Duff, G., Baranov, V., Sibul, H., Metspalu, A., Goldman, D., Martin, N., Duffy, D., Schmidtke, J., Estivill, X., O'Brien, S, J., and Dean, M., 1998, Dating the origin of the *CCR5-delta32* AIDS-resistance allele by the coalescence of haplotypes, *Am. J. Hum. Genet.* **62:**1507–1515.

Tajima, F., 1983, Evolutionary relationship of DNA sequences in finite populations, *Genetics* **105:**437–460.
Tajima, F., 1989, Statistical method for testing the neutral mutation hypothesis by DNA polymorphism, *Genetics* **123:**585–595.
Tishkoff, S. A., Dietzsch, E., Speed, W., Pakstis, A. J., Kidd, J. R., Cheung, K., Bonne-Tamir, B., Santachiara-Benerecetti, A. S., Moral, P., Krings, M., Paabo, S., Watson, E., Risch, N., Jenkins, T., and Kidd, K. K., 1996, Global patterns of linkage disequilibrium at the CD4 locus and modern human origins, *Science* **271:**1380–1387.
Watt, W. B., 1994, Allozymes in evolutionary genetics: Self-imposed burden or extraordinary tool? *Genetics* **136:**11–16.
Wayne, M., and Kreitman, M., 1996, Reduced variation at *concertina*, a heterochromatic locus in *Drosophila*, *Genet. Res.* **68:**101–108.

# 8

# Evolutionary Genetics of Primate Color Vision

## Recent Progress and Potential Limits to Knowledge

WEN-HSIUNG LI, STEPHANE BOISSINOT, YING TAN, SONG-KUN SHYUE, and DAVID HEWETT-EMMETT

## INTRODUCTION

Color vision has intrigued philosophers and scientists since Plato. It is truly a multidisciplinary subject involving physics, psychology, neuroscience, ophthalmology, genetics, evolution, molecular biology, biochemistry, and so forth. While past progress was made mainly from studies in physics, psychology, and vision science, recent progress has been mainly due to the advent of molecular and biochemical techniques such as gene cloning, DNA sequencing, site-directed mutagenesis, and biochemical assays of pigment proteins. Thanks to new techniques and traditional means, we now have a much better knowledge of the various genetic systems of color vision in primates, the origins and evolution of these systems, and the critical amino acid residues responsible for spectral sensitivity differences among primate

WEN-HSIUNG LI • Department of Ecology and Evolution, University of Chicago, Chicago, Illinois 60637. STEPHANE BOISSINOT • Laboratory of Molecular and Cellular Biology, National Institute of Diabetes and Digestive and Kidney Diseases, National Institutes of Health, Bethesda, Maryland 20892-0830. YING TAN and DAVID HEWETT-EMMETT • Human Genetics Center, University of Texas, Houston, Texas 77225. SONG-KUN SHYUE • Institute of Biomedical Sciences, Academia Sinica, Taipei 11529, Taiwan.

*Evolutionary Biology, Volume 32*, edited by Michael T. Clegg *et al.*
Kluwer Academic / Plenum Publishers, New York, 2000.

photopigments. These topics will be reviewed in this chapter. In addition, we shall discuss the role of natural selection in the evolution of primate color vision systems and also how gene conversion has greatly complicated the inference of the evolutionary history of the systems. We intend to show that primate color vision provides an excellent example of how misconceptions in a subject can be corrected and dramatic progress made by developing new approaches or collecting new data. The great progress notwithstanding, there are many potential limits to having a detailed knowledge of certain aspects of the field; some of these limitations will be discussed at the end of the chapter.

Before going into the subject matter of this review, it is useful to have a sketch of the primate phylogenetic relationships (Fig. 1). The primates are traditionally classified into the simians (higher primates) and the prosimians. The prosimians are divided into two groups: one group includes the lemurs, bush babies, lorises, and so on, and the other group consists of tarsiers. The tarsiers are now believed to be more closely related to the simians than to the other prosimians; that is, the prosimians are paraphyletic. The simians consist of the New World monkeys (NWMs) and the Old World primates. The latter includes *Homo sapiens* (humans), the apes, and the Old World monkeys (OWMs). The NWMs are often classified into two families, Cebidae and Atelidae; alternatively, Atelidae is further divided into two families, Atelidae and Pitheciidae (Schneider *et al.*, 1993, 1996). For a more

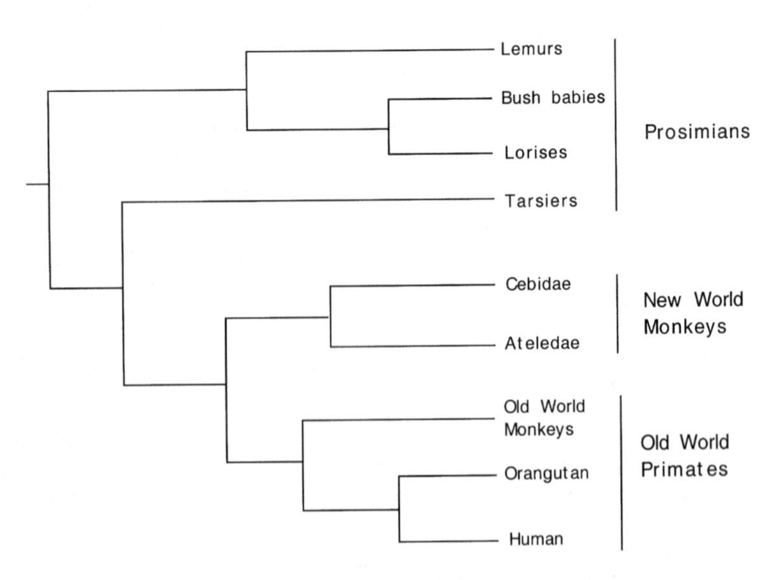


FIG. 1. Schematic representation of the primate phylogeny.

detailed description of primate classifications and phylogenetic relationships, see Goodman *et al.* (1998).

## COLOR VISION SYSTEMS IN PRIMATES

Prior to the molecular investigation of color vision, it was already known that humans, apes, and OWMs are trichromatic because they possess three types of pigments: the so-called blue (short wavelength), green (middle wavelength), and red (long wavelength) pigments. The blue pigment (or opsin, the apoprotein of a pigment) is encoded by an autosomal gene, whereas the green and red opsins are encoded by two X-linked duplicate genes. The cloning and sequencing of the human visual pigment genes (Nathans *et al.*, 1986a,b) have greatly facilitated the study of molecular genetics and evolution of color vision. It is now known that the red and 1–5 green opsin genes are tightly linked on the X-chromosome with the red opsin gene on the 5′ end of the tandem array. [However, since all the green opsin genes are almost identical and since only the 5′ most green gene appears to be expressed (Winderickx *et al.*, 1992), we shall not be concerned with the multiplicity of green genes.] The red and green opsin genes contain six exons (with a total of 364 codons), whereas the blue opsin gene contains only five exons (with 348 codons). Further, sequencing studies revealed that the high frequencies of green–red color blindness in human populations are at least due in part to recombination or gene conversion between the red and green opsin genes (e.g., see Nathans *et al.*, 1986b; Deeb *et al.*, 1992). Note that in addition to the three color photopigments, which are expressed in cone cells (cones), primates, like other animals, also have a rhodopsin, which is responsible for dim light vision and is expressed in rod cells (rods).

As it had long been firmly believed that all NWMs have only one X-linked opsin gene, it came as a big surprise when it was found, by a combination of psychophysical and molecular genetic analyses, that howler monkeys (*Alouatta* sp.) possess two X-linked opsin genes with spectral sensitivity peaks similar to those of human red and green pigments and are trichromatic (Jacobs *et al.*, 1996a; Boissinot *et al.*, 1997; Hunt *et al.*, 1998). However, all other NWMs studied possess only one X-linked opsin locus (see Jacobs, 1996; Boissinot *et al.*, 1997).

Prior to the use of molecular techniques, it also was known that some NWMs such as the squirrel monkey and the marmoset have three high-frequency alleles at the single X-linked opsin locus and that the spectral sensitivity maxima ($\lambda_{max}$) of these alleles are within those delimited by the human red and green pigments (e.g., Jacobs, 1984; Mollon *et al.*, 1984; Travis *et al.*, 1988; Jacobs *et al.*, 1993). Because of the existence of this triallelic system, heterozygous NWM females are trichromatic, though males and homozygous females are dichromatic. Using polymerase chain reaction (PCR) and DNA sequencing techniques, Shyue *et al.* (1998) and Boissinot *et al.* (1998) found that the triallelic system also exists in each of the three additional species studied: the capuchin, the tamarin, and the saki monkey. The capuchin and tamarin belong to the same family (Cebidae) as do the squirrel monkey and marmoset, but the saki monkey belongs to the other NWM family, Atelidae (or Pitheciidae, see above) and is one of the NWM species most divergent from the squirrel monkey and marmoset. Thus, the triallelic system appears to exist in the majority of NWMs.

Recent studies of color vision in prosimians have clarified several issues. In support of Mervis' (1974) investigation, Blakeslee and Jacobs (1985) found evidence that the ringtail lemur (*Lemur catta*), which is a diurnal prosimian, could make color discriminations, though the capacity was far from acute. However, Jacobs and Deegan (1993) found that ringtail lemurs and brown lemurs (also diurnal) have only a single class of photopigment in the middle to long wavelengths (sensitivity peak at ~545 nm) and a short wavelength pigment with peak at ~437 nm. They therefore concluded that both of these diurnal prosimians are dichromatic and speculated that the limited ability of ringtail lemurs to discriminate colors might have resulted from the ability of lemurs to jointly utilize signals from cones and rods.

Bush babies are nocturnal prosimians, and in apparent accord with this lifestyle, the early literature contained repeated claims that the bush baby retina contains only rods but no cones (see the review by Deegan and Jacobs, 1996). Further, although Dodt (1967) found a peak at ~552 nm in the spectral sensitivity curve determined by electroretinogram (ERG) and although Ordy and Samorajski (1968) obtained a similar result, the possibility that the bush baby retina might contain some cones was not raised. Moreover, no cones were detected in a more recent study using microspectrophotometric measurements of the absorbance properties of individual photoreceptors (Petry and Harosi, 1990). However, labeling the photoreceptors in *Galago garnetti* with cone-specific antibodies, Wikler and Rakie (1990) estimated the proportion of cones in the bush baby retina to be from ~1 to ~3%, depending on the retina location. Recently, using ERG flicker photometry, Deegan and Jacobs (1996) found that the cones of the thick-tailed bush baby (*Otolemur crassicaudatus*, formerly *Galago crassicaudatus*) contains a single type of photopigment with peak at ~545 nm.

As color vision presumably is of no use to a strictly nocturnal animal, it has been commonly thought that color pigment genes in an animal with a long history of nocturnal life should have evolved rapidly and would become degenerate or nonfunctional because of relaxation in their functional constraints. In fact, the blue (short wavelength) opsin gene in the bush baby (*O. crassicaudatus*) has accumulated deleterious mutations and become nonfunctional (Jacobs *et al.*, 1996b; Jacobs, 1996). However, unexpectedly, sequencing work revealed that the middle wavelength opsin gene in both *Galago senegalensis* and *O. garnettii*, which is closely related to *O. crassicaudatus*, has been well conserved (Zhou *et al.*, 1997); in fact, it has been even better conserved than the X-linked pigment genes in higher primates. It is possible that this opsin in combination with rhodopsin can provide the bush baby with a wider light spectrum at dusk, during which the animal is active, than can rhodopsin alone (Deegan and Jacobs, 1996). Moreover, since rhodopsin is shut off by daylight, the X-linked opsin might be the only functional opsin for the animal during daylight. Although bush babies usually are not active in the daytime, occasionally they will need to move (e.g., in order to escape predation), and thus need to use the X-linked opsin. Furthermore, it has been suggested that this opsin gene might play a role in the circadian rhythm of mammals (Nei *et al.*, 1997).

Since our main concern is the transition from dichromatic to trichromatic vision and since the blue opsin locus was not involved in this transition, this locus will not be discussed further.

## CRITICAL AMINO ACID RESIDUES FOR SPECTRAL TUNING

The spectral sensitivity peak ($\lambda_{max}$) is the most commonly used quantity for characterizing the phenotype of a pigment. This quantity is usually estimated by ERG or microspectrophotometry. Since the exact $\lambda_{max}$ is very difficult to determine, only approximate values are obtained. For example, the $\lambda_{max}$ values for the three alleles in squirrel monkeys were previously given as 538, 551, and 561 nm, respectively (Jacobs and Neitz, 1987), but were estimated to be 535, 550, and 562 nm in a recent reanalysis of previous data (Jacobs, 1996). The $\lambda_{max}$ values for the three alleles in the marmoset and tamarin are approximately 543, 556, and 562 nm (Jacobs, 1996). These alleles are commonly denoted as P535, P543, P550, P556, and P562. The most common green and red pigments in humans have the $\lambda_{max}$ values of ~530 and ~562 nm and are denoted as P530 and P562, respectively; a red pigment variant with alanine (Ala) instead of serine (Ser) at position 180 has a $\lambda_{max}$ of ~556 nm. Note that two pigments with the same $\lambda_{max}$ value (e.g., human P562 and tamarin P562) may have different origins and different amino acid sequences.

There has been much interest in knowing the amino acid residue sites that are involved in spectral tuning. One way to study this problem is to compare the amino acid sequences of closely related pigments with known $\lambda_{max}$ values. The opsins, as members of the superfamily of G-protein-coupled receptors, have seven transmembrane domains (Fig. 2). It has been hypothesized that only amino acid changes that are inside the transmembrane domains and involve a hydroxyl group (–OH) (i.e., Tyr, Ser, and Thr) can affect the spectral sensitivity of a pigment (see Nathans *et al.*, 1986a). Among the 15 amino acid differences between human red and green pigments (Fig. 2), only seven changes satisfy these two conditions (Table I). Since tamarin P562 and P556 differ at position 180, tamarin P556 and squirrel monkey P547 differ at position 277, and tamarin P556 and P541 differ at position 285 (Table I; the $\lambda_{max}$ values are old estimates), Neitz *et al.* (1991) estimated that the Ala–Ser, Ala–Thr, and Phe–Tyr changes at positions 180, 277, and 285 cause red shifts (increases in $\lambda_{max}$) of 6, 9, and 15 nm, respectively. These estimates were corroborated by a regression analysis of the

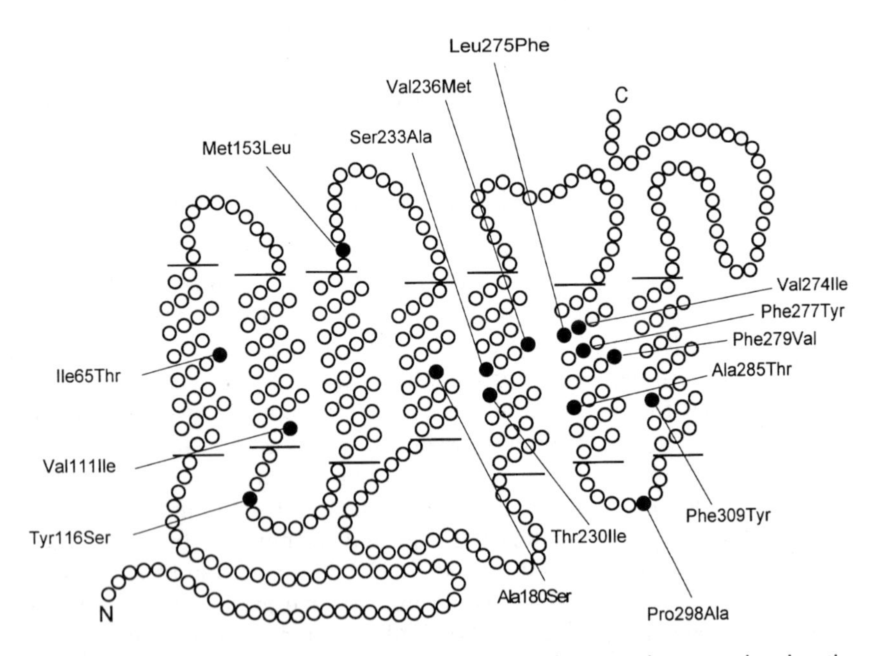


FIG. 2. Transmembrane model of the human red and green pigments showing the locations of amino acid differences (solid circles) and identities (open circles). The 15 amino acid differences are highlighted by solid circles in the figure. For each difference the amino acid residue in the green pigment is first followed by the position in the amino acid sequence and then by the amino acid in the red pigment.

eight primate opsin sequences then available. So, the three changes were assumed to be sufficient to explain the 31-nm difference between human red and green pigments (Table I).

Another way to study this problem is to introduce mutations either singly or in combination in an opsin cDNA by site-directed mutagenesis, express the mutant cDNA in animal cells, and measure the spectral sensitivity of each mutant sequence by spectrophotometry. Using this method, Asenjo *et al.* (1994) examined all the 15 amino acid differences between human red and green pigments (Fig. 2) and concluded that the spectral difference between the two pigments are determined by seven and only seven amino acid residues, which are, from the green pigment to the red pigment, Tyr116Ser, Ala180Ser, Thr230Ile, Ser233Ala, Phe277Tyr, Ala285Thr, and Phe309Tyr (Table II). Their study suggested some background effect: the spectral sensitivity effect of an amino acid change is weaker when the change is introduced into the green (middle wavelength) pigment than into the red (long wavelength) pigment, apparently because the latter has a longer wavelength (i.e., a higher $\lambda_{max}$). For example, at position 180 the replacement of Ala by Ser in the green pigment caused only an ~2-nm red shift, whereas the replacement of Ser by Ala in the red pigment caused an ~7-nm blue shift (decrease). According to their estimates, Ala180Ser, Phe277Tyr, and Ala285Thr caused red shifts of 2–7, 6–10, and 10–16 nm,

TABLE I. Comparison of Primate Opsin Sequences[a]

| | | Amino acid position | | | | | | |
|---|---|---|---|---|---|---|---|---|
| Subject | Spectral peak | 180 | 230 | 233 | 277 | 285 | 296 | 309 |
| Human | | | | | | | | |
| Red | 561 nm | Ser | Ile | Ala | Tyr | Thr | Ala | Tyr |
| Green | 530 nm | Ala | Thr | Ser | Phe | Ala | Pro | Phe |
| | Diff = 31 nm | | | | | | | |
| Monkeys | | | | | | | | |
| Tamarin | 562 nm | Ser[b] | Ile | Ala | Tyr | Thr | Ala | Tyr |
| Tamarin | 556 nm | Ala | Ile | Ala | Tyr | Thr | Ala | Tyr |
| | Diff = 6 nm | | | | | | | |
| Tamarin | 556 nm | Ala | Ile | Ala | Tyr | Thr | Ala | Tyr |
| Tamarin | 541 nm | Ala | Ile | Ala | Tyr | Ala | Ala | Tyr |
| | Diff = 15 nm | | | | | | | |
| Tamarin | 556 nm | Ala | Ile | Ala | Tyr | Thr | Ala | Tyr |
| Squirrel monkey | 547 nm | Ala | Ile | Ala | Phe | Thr | Ala | Tyr |
| | Diff = 9 nm | | | | | | | |

[a] From Neitz *et al.* (1991).
[b] The amino acids that differ between the two compared opsins are underlined.

TABLE II. Amino Acid Differences between Human Red and Green Opsins and Estimates of Their Effects on Spectral Shifts[a]

| | Position | | | | | | | | | | | | | | |
|---|---|---|---|---|---|---|---|---|---|---|---|---|---|---|---|
| | 65 | 111 | 116 | 153 | 180 | 230 | 233 | 236 | 274 | 275 | 277 | 279 | 285 | 298 | 309 |
| Human green | Ile | Val | Tyr | Met | Ala | Thr | Ser | Val | Val | Leu | Phe | Phe | Ala | Pro | Phe |
| Human red | Thr | Ile | Ser | Leu | Ser | Ile | Ala | Met | Ile | Phe | Tyr | Val | Thr | Ala | Tyr |
| Spectral shift (nm) | 0 | 0 | 0–4 | 0 | 2–7 | 0–4 | 0–3 | 0 | 0 | 0 | 6–10 | 0 | 10–16 | 0 | 0–3 |

[a]Data from Asenjo *et al.* (1994).

respectively, depending on the pigment sequence in which the mutation was introduced. These estimates were broadly in agreement with the corresponding estimates of ~5, 7, and 14 nm by Merbs and Nathans (1992, 1993), who introduced the mutations into the red pigment with either Ala or Ser at position 180. Asenjo *et al.* (1994) further obtained estimates of red shifts of 0–4, 0–4, 0–3, and 0–3 nm for Tyr116Ser, Thr230Ile, Ser233Ala, and Phe309Tyr, respectively, depending on whether the pigment into which the mutation was introduced was more similar to the human green or red pigment. However, Merbs and Nathans (1993) found the changes at positions 230, 233, and 309 to cause shifts of 1 nm or less when they were introduced into the red pigment with Ala at position 180. The latter estimates are smaller than the former (see discussion below).

Shyue *et al.* (1998) sequenced the three alleles at the X-linked opsin locus in the squirrel monkey, capuchin, marmoset, and tamarin. Comparing these sequences and assuming that Ala180Ser and Ala285Thr cause red shifts of 5 and 8 nm, respectively, they obtained estimates of −2, −1, and 15 nm shifts for Ile229Phe, Gly233Ser, and Ala285Thr, and found no discernable effect for His116Tyr, Val275Met, Leu275Met, and Ala276Thr (Table III).

TABLE III. Differences at Amino Acid Positions That May Be Involved in the Spectral Tuning of the Middle- and Long-Wavelength Visual Pigments of the Marmoset, Tamarin, Squirrel Monkey, and Capuchin[a]

| | Amino acid position | | | | | | | |
|---|---|---|---|---|---|---|---|---|
| Gene | 116 | 180 | 229 | 233 | 275 | 276 | 277 | 285 |
| Marmoset P562 | H | **S** | **F** | G | V | A | **Y** | **T** |
| Tamarin P562 | H | **S** | **F** | G | V | A | **Y** | **T** |
| Squirrel monkey P562 | **Y** | **S** | **F** | G | **M** | A | **Y** | **T** |
| Capuchin P562 | **Y** | **S** | **F** | G | **M** | A | **Y** | **T** |
| Marmoset P556 | **Y** | A | **F** | **S** | A | A | **Y** | **T** |
| Tamarin P556 | H | A | **F** | **S** | A | A | **Y** | **T** |
| Squirrel monkey P550 | **Y** | A | I | **S** | L | **T** | F | **T** |
| Capuchin P550 | H | A | I | **S** | L | **T** | F | **T** |
| Marmoset P543 | **Y** | A | I | **S** | V | A | **Y** | A |
| Tamarin P543 | H | A | I | **S** | V | A | **Y** | A |
| Squirrel monkey P535 | **Y** | A | I | **S** | V | **T** | F | A |
| Capuchin P535 | H | A | I | **S** | V | **T** | F | A |
| Spectrum shift (nm)[b] | 0 | 5 | −2 | −1 | 0 | 0 | 8 | 15 |

[a] From Shyue *et al.* (1998).
[b] The spectral shift refers to the shift in the spectral peak from a no-boldfaced amino acid to the boldfaced amino acid.

In summary, positions 180, 277, and 285 are the major critical sites for spectral tuning among the higher primate X-linked pigments, and the changes Ala180Ser, Phe277Tyr, and Ala285Thr, respectively, cause shifts of 4–7, 6–10, and 10–16 nm in these primate pigment sequences (Tables II and III). Positions 230 and 309 are considerably less critical because according to Merbs and Nathans (1993), Ile230Thr and Tyr309Phe cause shifts of 1 nm or less, though Asenjo *et al.* (1994) obtained estimates of 0–4 and 0–3 nm, respectively. The same comment applies to position 233 because Shyue *et al.* (1998) estimated a shift of –1 nm for Gly233Ser (Table III), and Merbs and Nathans (1993) estimated a shift of –0.4 nm for Ala233Ser, though Asenjo *et al.* (1994) gave an estimate of –0–4 nm for Ala233Ser. The significance of position 116 is not certain. Shyue *et al.* (1998) found that the change His116Ser had no discernable effect on $\lambda_{max}$, regardless of whether it occurred in a middle or a long wavelength pigment (Table III). Also, no spectral effect was found for the change Tyr116Ser when it was introduced into a middle wavelength pigment (Asenjo *et al.*, 1994). However, when it was introduced into a long wavelength pigment a 3- to 4-nm shift was estimated (Asenjo *et al.*, 1994). One possible explanation for the discrepancy between His116Ser and Tyr116Ser is that the difference in molecular size (weight) between Tyr (181) and Ser (105) is considerably greater than that between His (155) and Ser. If Tyr116Ser indeed has a spectral effect under certain conditions, then it is an exception to the hypothesis that only changes within the transmembrane domains can have spectral effects, because position 116 is outside the transmembrane domains, though close to the boundary (Fig. 2). It also is interesting to note that both Tyr and Ser have an –OH group, so this substitution involves no change in –OH. Note also that the change Ile229Phe involves no change in –OH but may cause a shift of –2 nm (Table III). So the hypothesis that only changes that involve an amino acid with an –OH group can have an effect on spectral sensitivity may not be strictly true.

It is useful to note that position 116 is in exon 2, position 180 is in exon 3, positions 229, 230, and 233 are in exon 4, and positions 277, 285, and 309 are in exon 5. There is no critical amino acid residue in either exon 1 or exon 6.

The above residues are those that are involved in the spectral tuning of X-linked pigments in higher primates. Sites 185, 277, and 285 also have been found to be major critical sites in the spectral tuning of red–green vision in other vertebrates (Yokoyama and Yokoyama, 1990; Yokoyama and Radlwimmer, 1998). Additional critical residues have been found in other mammals. For example, the changes His197Tyr and Ala308Ser were estimated to cause 28- and 18-nm shifts in mouse, respectively (Sun *et al.*, 1997). For further details, see Yokoyama and Radlwimmer (1998).

## FREQUENT GENE CONVERSION BETWEEN X-LINKED OPSIN ALLELES OR GENES

The tight linkage and high similarity (~98%) between the red and green pigment genes in Old World primates provide a favorable condition for gene conversion to occur between them. This possibility was first noted by Balding *et al.* (1992) and Ibbotson *et al.* (1992) for exons 4 and 5 between the two genes in OWMs. Later, in a study of haplotype diversity in human red and green opsin genes, Winderickx *et al.* (1993) found evidence for frequent sequence exchange in exon 3 between the two genes. Deeb *et al.* (1994) and Reyniers *et al.* (1995) provided further evidence of frequent gene conversion between the two genes in humans and OWMs.

A remarkable example of gene conversion was found when introns 4 of human red and green opsin genes were completely sequenced (Shyue *et al.*, 1994). The two introns were identical, though a divergence of >8% was expected between them because the human red and green opsin genes arose from a duplication before the divergence of the OWM and human lineages (i.e., more than 25 million years ago). The individual studied by Shyue *et al.* was a European, but Zhou and Li (1996) found that this also was true for an Asian Indian (Table IV). Moreover, the intron 2 sequences

TABLE IV. Mean and Standard Error of the Number of Nucleotide Substitutions per 100 Sites in Intron 4 and in Exons 4 and 5 between the Red and Green Pigment Genes[a]

| | Intron 4 | Exon 4 + exon 5 | |
|---|---|---|---|
| Species and genes[b] | $K^c$ | $K_S{}^d$ | $K_A{}^d$ |
| Human G/Human R | 0.0 ± 0.0 | 8.1 ± 4.3 | 5.8 ± 2.0 |
| Chimpanzee G/Chimpanzee R | 0.3 ± 0.1 | 6.6 ± 3.5 | 5.1 ± 1.8 |
| Baboon G/Baboon R | 0.9 ± 0.2 | 11.5 ± 5.1 | 5.1 ± 1.8 |
| Human G/Chimpanzee G | 1.1 ± 0.3 | 4.2 ± 3.2 | 0.0 ± 0.0 |
| Human R/Chimpanzee R | 1.0 ± 0.3 | 0.0 ± 0.0 | 0.7 ± 0.7 |
| Human G/Baboon G | 7.3 ± 0.7 | 5.8 ± 3.6 | 1.1 ± 0.8 |
| Human R/Baboon R | 7.1 ± 0.7 | 9.0 ± 4.3 | 1.3 ± 1.0 |
| Chimpanzee G/Baboon G | 7.8 ± 0.8 | 6.0 ± 3.1 | 1.1 ± 0.8 |
| Chimpanzee R/Baboon R | 7.5 ± 0.7 | 9.0 ± 4.3 | 0.7 ± 0.7 |

[a] From Zhou and Li (1996).
[b] G and R represent the green and red pigment genes, respectively.
[c] The *K* value is the number of substitutions per 100 sites and was computed by Kimura's two-parameter method (Kimura, 1980).
[d] The $K_S$ and $K_A$ values are the numbers of substitutions per 100 synonymous sites and per 100 nonsynonymous sites, respectively, and were computed by Li's methods (Li, 1993).

of the two genes differed by only 0.3% (Shyue *et al.*, 1994). The complete human red opsin gene and a large region (from the 3′ part of intron 3 to the 3′ flanking region) of the human green opsin gene have been sequenced, and a comparison reveals that all noncoding parts that are now available for comparison (the 3′ part of intron 3, intron 4, intron 5, and the 3′ flanking region) are identical or almost identical between the two genes (Table V), suggesting frequent gene conversion between the two genes (Zhao *et al.*, 1998). In contrast, the divergences in exons 4 and 5 are 3.8% and 4.7% (Table V). It is likely that gene conversion also has occurred in exons (see above). In fact, exons 1 and 6 of the red and green opsin genes have been completely homogenized (Table V) (Shyue *et al.*, 1994; Nathans *et al.*, 1996a). This observation is not surprising because both exons 1 and 6 contain no critical amino acid residue (see above). However, exons 2, 3, 4, and 5 each contain residues that are critical to the spectral differences between the red and green opsin peptides. A gene conversion event in any of these exons may reduce the spectral sensitivity differences between the two opsins, so it may be disadvantageous and eliminated from the population. This probably is the reason that exons 2, 3, 4, and 5 of the two genes have been maintained distinct.

Zhou and Li (1996) also found that the degree of divergence between the intron 4 sequences of the red and green opsin genes was only 0.3% in a chimpanzee and 0.9% in a baboon, indicating that gene conversion also

TABLE V. Mean and Standard Error of the Number of Nucleotide Substitution per 100 Sites between Human Red and Green Pigment Genes in Exons, Introns, and 3′ Flanking Sequences[a,b]

| | No. of differences/ sequence length (bp) | Noncoding region ($K$) | Coding region | |
|---|---|---|---|---|
| | | | $K_S$ | $K_A$ |
| Introns 3 | 0/506 | 0.0 ± 0.0 | | |
| Exon 4 | 5/166 | | 3.8 ± 3.0 | 3.3 ± 1.9 |
| Intron 4 | 1/1554 | 0.1 ± 0.1 | | |
| Exon 5 | 10/240 | | 4.7 ± 2.9 | 3.9 ± 1.5 |
| Intron 5 | 2/2282 | 0.1 ± 0.1 | | |
| Exon 6 | 0/108 | | 0.0 ± 0.0 | 0.0 ± 0.0 |
| 3′ Flanking | 2/1256 | 0.2 ± 0.1 | | |
| Exon 4, 5, and 6 | 15/514 | | 3.4 ± 1.6 | 3.0 ± 1.0 |

[a] From Zhao *et al.* (1998).
[b] The $K$ value was computed by Kimura's two-parameter method (Kimura, 1980), and the $K_S$ and $K_A$ values were computed by Li's method (Li, 1993). Only about one third of intron 3 was available for comparison.

has occurred in these two species (Table IV). In comparison, exons 4 and 5 have diverged more than 6% at synonymous sites between the red and green opsin genes in each of these two species and in human. When the synonymous divergences are used to infer the relationships among the human and baboon red and green opsin genes, human and baboon red opsin genes are clustered together and so are human and baboon green opsin genes (Fig. 3). This tree is in agreement with the view that the red and green opsin genes diverged before the divergence of the OWM and human lineages. On the other hand, when a tree is constructed from the intron 4 sequences, the two human genes are clustered together and so are the two baboon genes (Fig. 3). This example shows that gene conversion can drastically distort attempts to reconstruct the evolutionary history of genes.

Gene conversion also has been inferred to have occurred between alleles of the same locus. A very dramatic case was found between the P535 and P535c alleles in capuchins (Boissinot *et al.*, 1998); P535c is a minor allele, and P535, P550, and P562 are the three major alleles. The P535 and P535c alleles are characterized by the same amino acids at positions, 180, 277, and 285 (the three major critical positions; see above), and their exons 3 and 5 are identical (Fig. 4). Yet, they differ in exon 4 by 3.6% and in intron

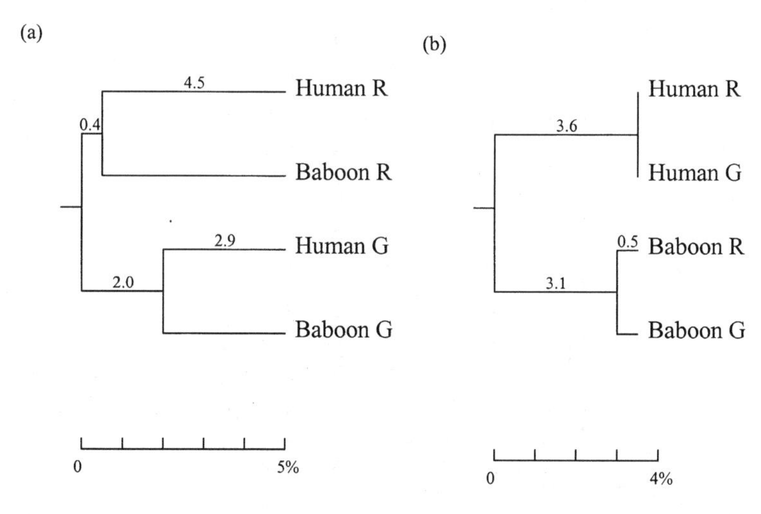


FIG. 3. (a) Tree inferred from exons 4 and 5 of the red (R) and green (G) pigment genes of human and baboon. (b) Tree inferred from intron 4 sequences of human and baboon red and green pigment genes. In each tree, the topology was inferred by the neighbor-joining method and the root was placed in between the two clusters. The branch lengths were computed by assuming equal rates in different branches. From Li (1997).

```
          EXON 3  EXON 4  INTRON 4                                                            EXON 5
          --      ------  -----------------------------------------------------------------  ------
                                                                        1111111111111111111
          55      666677                      11111122222222222223333334567990000002244446778889  888888
          35      589904   2233333333334491114590223444555580225596490782236992333482082666  022235
          85      256784  68901234567890193894630122149012796784929062731570481749273768237  136803

P562      TG      TTTGGG  TC-----------ACATCAGCTGGGGGT---GGACTGTTG-ACGC-TG-GGGCGAACAGCGGCGC  GAGAAA
                  ||||||  | ||||||||||| | |||||| | |||||||||||||||    ||||| ||||| ||||||  |
P535c     GA      TTTGGG  TA-----------CC-TCAGCTAGAGGT---GGACTGTTGAGA-C-TG-CGGCGATCAGCGGAAC  AGAGTG
          ||               |                    | |              |||      |    |            ||||||
P535      GA      CACAAT  CAGGGAAAGGGGGTGA-TGATCAAAAACCCATAGTCACCAAGAGTACTCC-AT-GTTCTAAACGG  AGAGTG
```

FIG. 4. Variable sites in exons 3, 4, and 5 and intron 4 of the P562, P535, and P535c alleles of the capuchin. The numbers at the top refer to the positions of the sites on the complete coding sequence (for exons 3, 4, and 5) and to the positions on an alignment of intron 4 sequences (available on request). The five positions in boldface are the critical amino acid residues, respectively, at positions 180, 229, 233, 277, and 285. From Boissinot *et al.* (1998).

4 by 2.0%. In contrast, P535c and P562 are identical in exon 4 and differ by only 0.4% in intron 4 (Fig. 4). A test by P. M. Sharp's modified method of Maynard Smith (1992) provides strong evidence ($P = 0.002$) that exon 4 and intron 4 of P535c have been converted by P562.

Allelic gene conversion also was indicated by the fact that the degree of divergence between alleles is not uniform along the intron 4 sequence, suggesting that some parts of intron 4 have been converted more recently than others (Boissinot *et al.*, 1998). For instance, in marmoset (Fig. 5), the divergence (~8.0%) in the first 600 base pairs (bp) of intron 4 is more than three times that in the rest of intron 4 (~2.5%) when P562 is compared with P543 or P556. In fact, the 8% divergence is even higher than the between-species divergence among the marmoset, tamarin, squirrel monkey, and capuchin (e.g., an average 5.1% divergence between the marmoset P543 and squirrel monkey P535 alleles at the bottom panel of Fig. 5), indicating that marmoset P562 and P543 (or P556) diverged before the divergence of these four species. Similarly, when P543 and P556 are compared, the first 400-bp portion is more than twice as divergent as the rest of intron 4 (~6.0% vs. ~2.5%). All three within-species comparisons in Fig. 5 are significant ($P < 0.001\%$, $0.01\%$, and $1\%$, respectively). This finding suggests that the 3′ portion of intron 4 was partially homogenized in each case by recent conversion events, whereas the 5′ portions were either not homogenized or were last homogenized by much earlier conversion events.

## ORIGINS OF COLOR VISION SYSTEMS IN HIGHER PRIMATES

An intriguing question is, "How did the various X-linked color vision systems in higher primates arise?" For example, were the red (P530) and green (P562) opsin genes in the Old World primates derived from two identical (or similar) alleles or from two alleles similar to the P535 and P562 alleles in NWMs? In the latter case, the two resultant duplicate genes, together with the blue opsin, would immediately confer trichromacy. Such an incorporation of two overdominant alleles into one chromosome has been proposed to be a possible advantage of gene duplication (Spofford, 1972), but no such example has yet been found. Another intriguing question is whether the triallelic systems in different NWM species have a single origin or multiple origins?

To resolve these and related issues, Shyue *et al.* (1995) sequenced exons 3, 4, and 5 and intron 4 of the three highly polymorphic alleles in the squirrel monkey and marmoset. Their data clearly indicated that the human red and green opsin genes have an origin independent of that of the triallelic

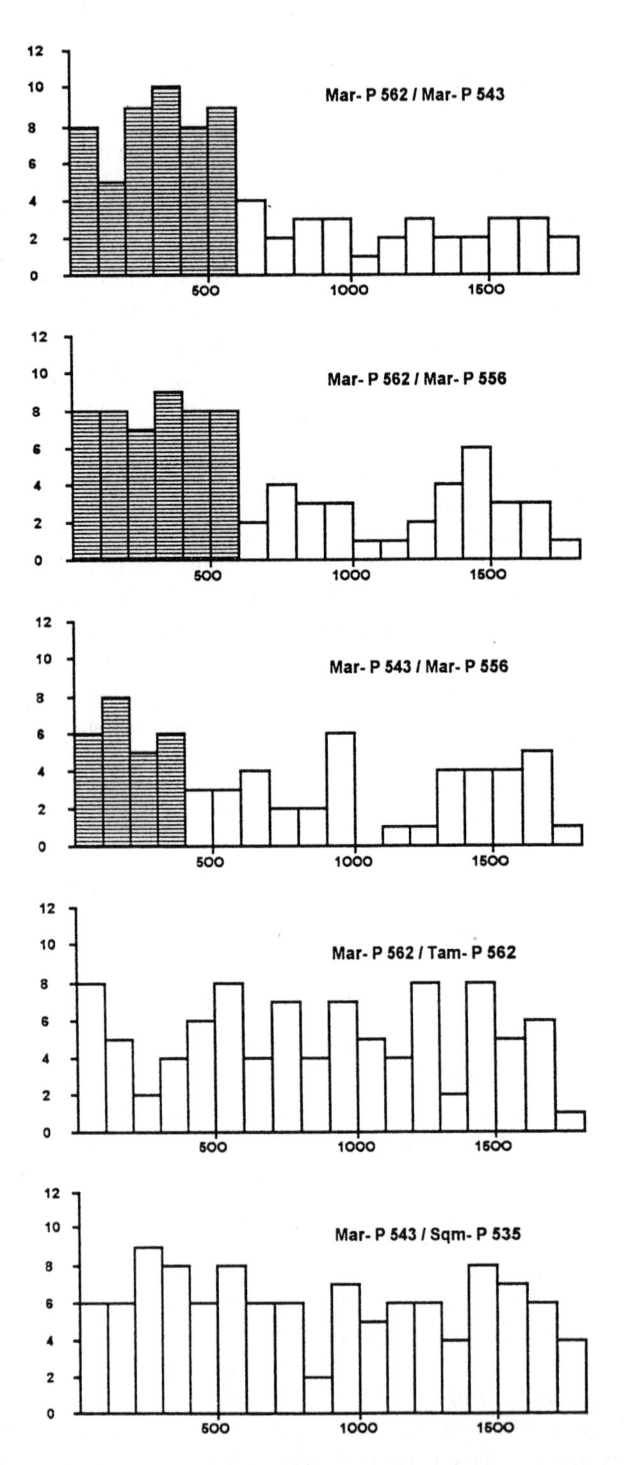


FIG. 5. Histograms showing pairwise variation in divergence as a function of position in intron 4. Horizontally each bar represents a 100-bp segment. Vertically each bar shows the number of nucleotide differences per segment. Three within-species (the top three figures) and two between-species comparisons are shown (Mar, marmoset; Tam, tamarin; Sqm, squirrel monkey). From Boissinot *et al.* (1998).

system in NWMs because an *Alu* repeat was present in the intron 4 sequences of all of the NWM alleles studied but absent from both human genes. This conclusion is further supported by new sequence data, which show that the *Alu* repeat in intron 4 is also present in the three alleles from the capuchin, tamarin, and saki monkey (Fig. 6a). In addition, the two howler monkey genes and the two human genes evidently have separated origins, because the former contain the *Alu* repeat while the latter do not. However, it remains possible that the human red and green opsin genes were derived from two different alleles that might have existed in the common ancestor of the Old World primates rather than from a single allele.

Shyue and co-workers' (1995) intron 4 sequence data favored the multiorigin hypothesis over the single-origin hypothesis because the three alleles in the squirrel monkey formed one monophyletic group and so did the three alleles in the marmoset. The monophyly of the alleles in each species is also found in Fig. 6a, which also includes the capuchin, tamarin, and saki monkey. This tree was inferred by the neighbor-joining method (Saitou and Nei, 1987). For the parsimony analysis of insertions and deletions (indels, Fig. 6a), the monophyly of the alleles in a species is supported by six indels in the marmoset, one indel in the tamarin, five indels in the squirrel monkey, and seven indels in the capuchin. Therefore, both the neighbor-joining tree and the indel analysis strongly support the multiorigin hypothesis; in fact, they imply five independent origins for the triallelic system, one in each species (Fig. 6a). Further, tree 6A implies that the two duplicate genes in the howler monkey had an origin independent of the triallelic system in the other NWMs.

However, the clustering of the alleles in each species is probably due to gene conversion. As described above, gene conversion often occurs between alleles in a population. Previously, it was thought that transferring of indels between alleles is unlikely to occur, so the clustering of alleles within each species was taken as evidence for an independent origin. However, the indel from positions 29 to 39 in intron 4 of capuchin P562 obviously has been transferred to P535c (Fig. 4). In Fig. 6a, the indels within each species also suggest indel transferring. For example, in tamarin, two indels are shared by P543 and P562, four by P562 and P556, and one by P543 and P556. It is highly unlikely that so many indels arose independently in different alleles. So, this pattern strongly suggests the transfer of indels between alleles by gene conversion, partially homogenizing the allelic sequences.

Although the evolutionary relationships among the alleles have apparently been drastically distorted by gene conversion, the branching order for the six NWMs in Fig. 6a is in agreement with the current view that the

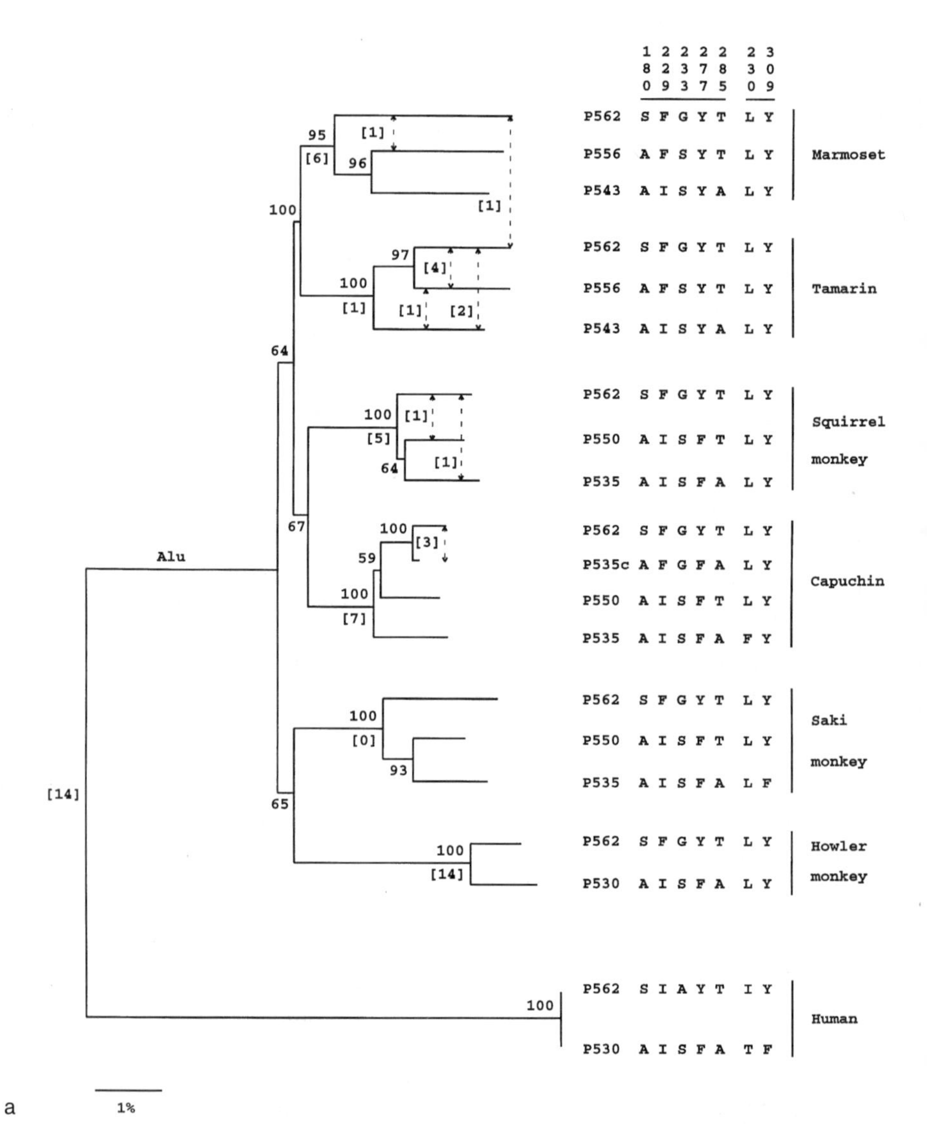


FIG. 6. (a) The neighbor-joining tree derived from intron 4 sequences. The distances were computed using Kimura's (1980) two-parameter method. The number at each node denotes the proportion of 500 bootstrap replicates that supported the subset of sequences. The number in brackets below a branch denotes the number of indels supporting the monophyly of the three alleles in each species. A dashed arrow with a number in brackets denotes the number of indels shared by two alleles. The seven vertical numbers at the upper right of the tree refer to the positions of the seven critical amino acid sites. (b) The neighbor-joining tree derived from exons 3, 4, and 5 sequences. The distances were computed using Li's (1993) method, with weights of 80% and 20% for nonsynonymous and synonymous substitutions, respectively. The number at each node denotes the percent of 500 bootstrap replicates that supported the subset of sequences. The galago sequences (Zhou *et al.*, 1997) were used to root the tree (CAP, capuchin; HUM, human; HOW, howler monkey; MAR, marmoset; SAK, saki monkey; SQM, squirrel monkey; TAM, tamarin). From Boissinot *et al.* (1998).

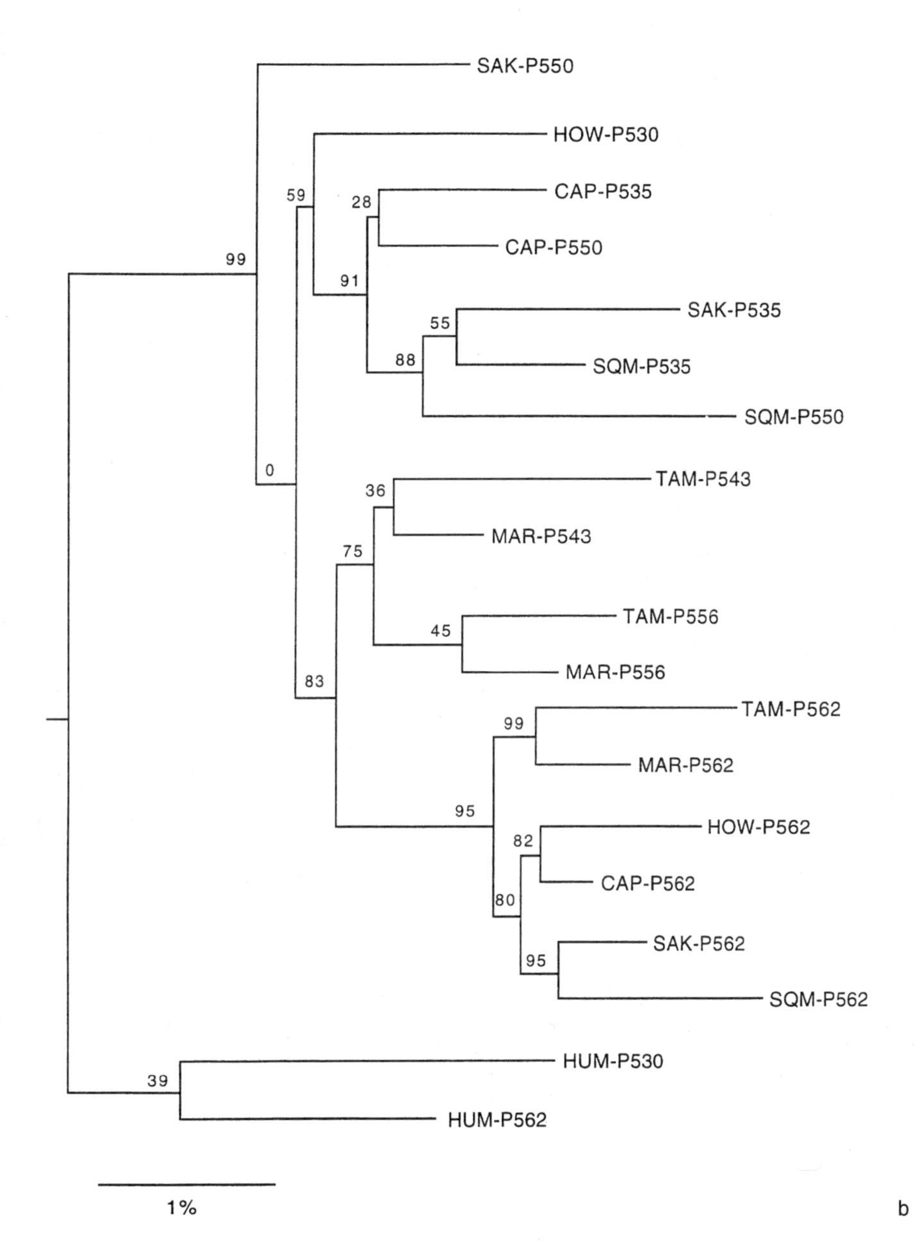


FIG. 6. (*Continued*)

marmoset–tamarin and squirrel monkey–capuchin clades belong to the same family (Cebidae) and that the saki monkey and howler monkey belong to the other family (Atelidae). When the *Alu* sequence, which is present in all NWM sequences but absent in the two human sequences, is included for inferring the phylogeny of the six NWM species, the clustering of the saki monkey with the howler monkey and that of the squirrel monkey with the capuchin are both strengthened (the bootstrap values become 94% and 88%).

Figure 6b shows the neighbor-joining tree based on the sequences of exons 3, 4, and 5. The clustering of the sequences is not in good agreement with the species tree; for example, saki monkey P562 (SAK-P562) is clustered with squirrel monkey P562 (SQM-P562) rather than with howler monkey P562 (HOW-P562). One reason for this can be because these exons also are not free from the effects of gene conversion. However, these exons should be less affected by gene conversion than introns because they contain critical sites for spectral tuning and a gene conversion involving a critical site may be eliminated from the population by natural selection. Thus, tree 6b should be more reliable than tree 6a for depicting the origin of the alleles. Clearly, it does not support the multiorigin hypothesis of tree 6A. In particular, the P562 alleles of the saki monkey, squirrel monkey, capuchin, marmoset, and tamarin are clustered, suggesting that these alleles were derived from a single origin. Moreover, howler monkey P562 belongs to this cluster, while howler monkey P530 is clustered with the P535 alleles of the saki monkey, squirrel monkey, and capuchin, suggesting that the two howler monkey duplicate genes were derived from a combination of the P562 and P535 alleles.

As gene conversion can drastically mislead phylogenetic analysis at noncritical sites for spectral tuning, the best data for inferring the evolutionary history of these X-linked color vision alleles and duplicate genes are probably the amino acid changes at the critical sites. Sites 180, 277, and 285 are the major critical sites, causing $\lambda_{max}$ shifts of ~5, 8, and 15 nm, respectively, whereas sites 116, 229, 230, 233, and 309 have minor spectral tuning effects, causing shifts of <3 nm. Sites 229 and 233 show a substitution pattern highly consistent with the three major sites (Fig. 6a). However, sites 230 and 309 show very minor variation among sequences, so they are not informative for our purpose and will not be considered further. Site 116 is in exon 2, which is not under study. In terms of parsimony, tree 6a is highly implausible because it requires many parallel amino acid changes at the five critical sites considered. For example, at position 180, at least seven parallel changes between serine (S) and alanine (A) are required to explain the differences among alleles and genes at this site. When all five critical sites are considered together, the minimum number of substitutions required is 37 or 38, depending on whether one assumes that the amino acids at sites

180, 229, 233, 277, and 285 in the common ancestor of all higher primates were SFGYT (i.e., as in marmoset P562), or AISFA (as in saki monkey P535, or AISYA as in marmoset P543). (Capuchin P535c is a minor allele and, for simplicity, is not considered.) In comparison, Fig. 7 requires only 11 or 12 amino acid substitutions in the higher primate sequences, assuming that the ancestral amino acids were AISFA (as in saki monkey, squirrel monkey, and capuchin P535) or AISYA (as in marmoset and tamarin P543). Both scenarios require 13 substitutions for the entire tree (i.e., including also the two galago sequences) and are the most parsimonious scenarios subject to the constraint of the current knowledge of the phylogeny of these primates (i.e., the species phylogeny of tree 6a). A slightly more parsimonious scenario (one fewer change) is to assume that the marmoset–tamarin P556 was derived from the common ancestor of marmoset and tamarin P562. However, this scenario requires that the P550 allele has become lost in the common ancestor of the marmoset and tamarin (or in both species), even though it has persisted in the saki monkey, squirrel monkey, and capuchin.

In summary, since Fig. 7 requires less than one third of the number of critical amino acid changes required by Fig. 6a (11 or 12 vs. 37 or 38), the single-origin hypothesis is far more plausible than the multiorigin hypothesis suggested by the intron 4 sequences. Moreover, a parsimony analysis with the galago sequences as outgroups suggests that the X-linked opsin alleles and duplicate opsin genes in the higher primates (including humans) were derived from a middle wavelength (green) opsin gene similar to either P543 (Jacobs, 1993) or P535 (Winderickx *et al.*, 1993) but not from a long wavelength (red) opsin gene as inferred without the galago sequences (Nei *et al.*, 1997). Finally, the two howler monkey duplicate genes (P530 and P562) have a separate origin from the human red and green pigment genes, supporting the view of Jacobs *et al.* (1996a), Boissinot *et al.* (1997), and Hunt *et al.* (1998). They evidently were derived from a combination of the P535 and P562 alleles. If this suggestion is substantiated by further data, it will provide the first example where the recombination of two overdominant alleles into one chromosome provides the basis for the selective advantage of gene duplication (Spofford, 1972).

## EVOLUTIONARY MECHANISMS

It is now clear that trichromacy has arisen in higher primates in at least three different ways: (1) gene duplication and subsequent divergence in the common ancestor of the Old World primates, (2) a triallelic system in the NWMs except the howler monkeys, and (3) incorporation of two different

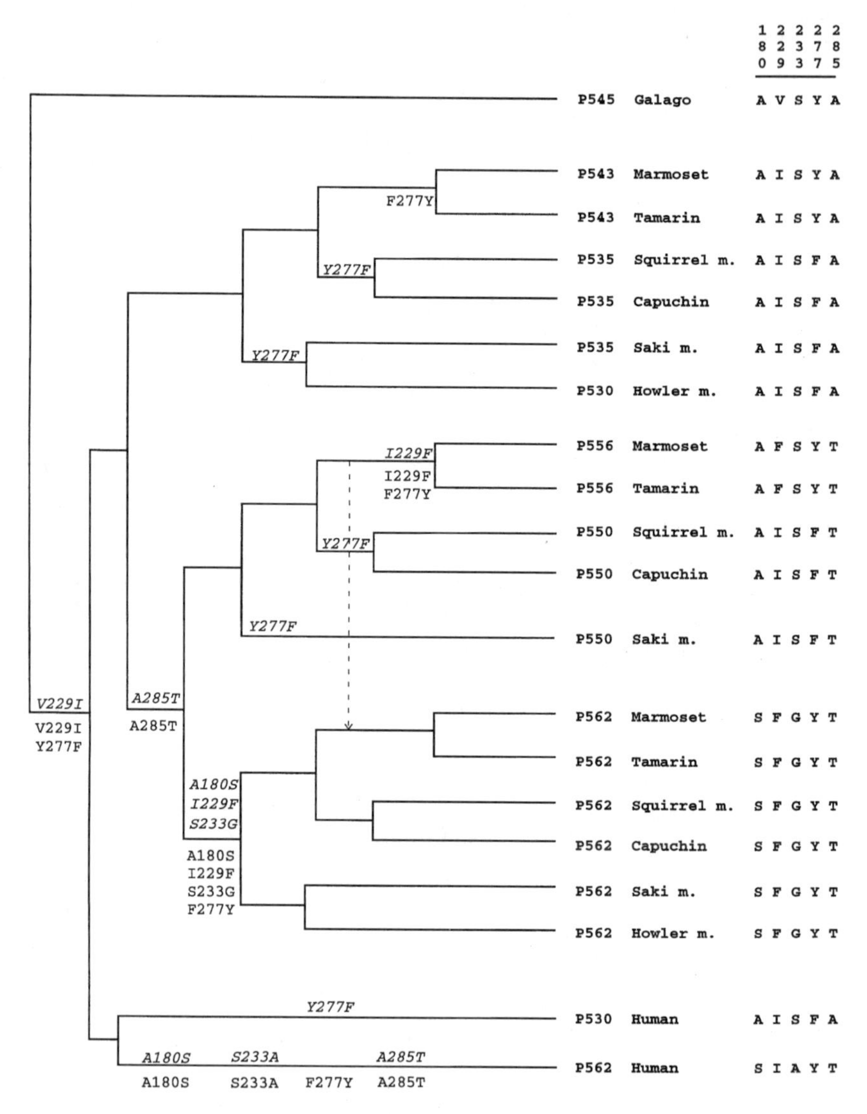


FIG. 7. The tree that requires the minimal number of amino acid changes at the five critical sites under the constraint of the species phylogeny in Fig. 6a. Substitutions along branches are indicated as the ancestral amino acid followed by its position on the complete coding sequence and then followed by the new amino acid. Two equally parsimonious pathways are shown, one above branches (italic) and one below branches. A shorter tree (12 instead of 13 steps) is obtained if the marmoset–tamarin P556 allele has a recent origin and was derived from the P562 allele, as indicated by the arrow. From Boissinot *et al.* (1998).

alleles into one chromosome in the common ancestor of howler monkeys. These repeated occurrences of trichromacy point to the selective advantage of trichromacy over dichromacy. The advantage is commonly believed to be for detecting yellow or red fruits against a green foliage background, because dichromatic vision cannot distinguish between red and green colors.

As noted above, the single-origin hypothesis for the triallelic system appears to be far more plausible than the multiorigin hypothesis. Note that the single-origin hypothesis implies that the triallelic system has persisted in these NWMs for more than 20 million years, because the divergence of the howler monkey lineage and the squirrel monkey–marmoset lineage has been estimated to be about 20 million years (Schneider *et al.*, 1993, 1996). The antiquity of the system strongly suggests balancing selection for the maintenance of the system, because without balancing selection one or two of the three alleles would have become lost in a relatively short time (Kimura and Ohta, 1969; Takahata and Nei, 1990). Such an antiquity of alleles have been known only in major histocompatibility complex and immunoglobulin genes (see Klein *et al.*, 1993). Note that the existence of three rather than two polymorphic alleles at the X-linked opsin locus increases the chance of being heterozygous, and thus of being trichromatic for a female NWM. However, if the triallelic system has multiple origins, then the repeated occurrences and the numerous parallel amino acid substitutions required to explain the sequence differences within and between species (see Fig. 6a) also suggest positive Darwinian selection. Actually, even the most parsimonious tree (Fig. 7) requires several parallel substitutions at some critical sites (e.g., at least four F→Y changes at site 277). Thus, regardless of whether the single-origin or the multiorigin hypothesis is true, the triallelic system probably has been maintained by natural selection.

## CONCLUDING REMARKS

It is clear from this review that much progress has been made in the evolutionary genetics of primate color vision. It is fairly certain that the origin of the two X-linked opsin genes in the Old World primates is independent of that of the triallelic system at the single X-linked opsin locus in NWMs. Moreover, there have been several changes of view. First, contrary to the long traditional belief that all NWMs have only one X-linked opsin gene, howler monkeys in fact have two X-linked opsin genes. That is, at least one NWM genus (*Alouatta*) has used gene duplication instead of allelic polymorphism to achieve trichromacy. Second, the ringtail lemur is dichromatic, not trichromatic. Third, the view that the bush babies have no cones is wrong; although their blue pigment gene is nonfunctional, their X-linked color pigment gene is functional and in fact has been well conserved. Fourth, indel transferring between X-linked opsin alleles or genes was thought to be unlikely to occur, but available sequence data strongly indicate that it has occurred several times in the NWMs studied.

The hypothesis that the two howler monkey X-linked opsin genes were derived from the incorporation of two alleles similar to present-day P535 and P562 remains to be substantiated by further data. Although this hypothesis is supported by the parsimony analysis of the critical amino acids (Fig. 7), it is not completely consistent with the neighbor-joining analysis of the entire sequences of the exons 3, 4, and 5 (Fig. 6b). The difficulty is mainly due to gene conversion between alleles or genes. None of the four other NWM species studied are closely related to the howler monkeys, so there has been considerable time for gene conversion to obscure the true evolutionary relationships. Currently we are studying two species of *Ateles* (*A. chamek* and *A. geoffroyi*) because *Ateles* is one of the genera that are most closely related to *Alouatta*, the howler monkeys. One limitation here is that it is not easy to obtain sufficiently large numbers of DNA samples for screening the alleles in each species.

In the past 5 years we have gained much better understanding of the evolutionary history of the triallelic systems in the NWMs. It is now clear that the system exists in the majority of NWMs. Also, although the multiorigin hypothesis was favored previously, the addition of sequence data from three new species reveals that the single-origin hypothesis is much more plausible. The sequence data to be obtained from *A. chamek* and *A. geoffroyi* will provide further insights into this issue. The main difficulty here again is that frequent allelic gene conversion has drastically distorted the evolutionary history of the triallelic system. Indeed, although it now appears that the system arose after the OWM–NWM split and in the common ancestor of the extant NWMs, the exact age of the system probably is not knowable because it has been almost completely obscured by gene conversion. Similarly, the exact date of the duplication for the two howler monkey X-linked opsin genes is not knowable.

In tracing evolutionary history, the major difficulty created by gene conversion is that once a sequence is converted by another sequence, its past history is completely lost. So, for example, it is likely that we can never know how many gene conversion events have occurred between the intron 4 sequences of human red and green pigment genes and when the last gene conversion event occurred between the two sequences.

The importance of gene conversion in population genetics and evolution has not been duly recognized. This is because a gene conversion event is very difficult to detect if the divergence between the two sequences involved is small, simply due to lack of statistical power. As a consequence, gene conversion has not been thought to be an important factor in population genetics and evolution. We and others have been able to detect frequent gene conversion events between NWM opsin alleles (or genes) because the divergences between these alleles are unusually large. In most other genes the degrees of divergence between alleles in a population are likely to be small and gene conversion events are unlikely to be noticed. Indeed, as noncritical regions of the alleles in a population can be readily homogenized by allelic gene conversion, even ancient alleles, which may have been maintained by balancing selection, may not become very divergent. In this case, sequence data may lead to erroneous phylogenetic inferences and to a severe underestimation of the antiquity of the alleles. In the case of the triallelic system in NWMs, we can appreciate the antiquity of the system because it has been found in many species. If the system were now persisting in only a single species, or if only one NWM species had survived, the triallelic system would be mistaken as a new polymorphism and its antiquity would be grossly underestimated.

Finally, another limitation to a detailed knowledge of the evolutionary genetics of color vision is the difficulty in obtaining accurate measurements of the $\lambda_{max}$ value of a pigment. As only approximate $\lambda_{max}$ values are obtained, the exact effect of an amino acid change on the spectral sensitivity of a pigment is not known. Consequently, it is difficult to know in detail how the effect of an amino acid change varies with the sequence background and how strong interactions between changes at different sites are. It is possible that more precise instruments or measurements will be developed in the future, but it is unlikely that an extensive study of this problem will ever again be conducted because of limitations in financial resources.

### Acknowledgments

This study was supported by National Institutes of Health grants. We thank the two reviewers for insightful suggestions.

## REFERENCES

Asenjo, A. B., Rim, J., and Oprian, D. D., 1994, Molecular determinants of human red/green color discrimination, *Neuron* **12:**1131–1138.

Balding, D. J., Nichols, R. A., and Hunt, D. M., 1992, Detecting gene conversion: Primate visual pigment genes, *Proc. R. Soc. London B* **249:**275–280.

Blakeslee, B., and Jacobs, G. H., 1985, Color vision in the ring-tailed lemur (*Lemur catta*), *Brain Behav. Evol.* **26:**154–166.

Boissinot, S., Zhou, Y.-H., Qiu, L., Dulai, K. S., Neiswanger, K., Schneider, H., Sampaio, I., Hunt, D. M., Hewett-Emmett, D., and Li, W.-H., 1997, Origin and molecular evolution of the X-linked duplicate color vision genes in howler monkeys, *Zool. Stud.* **36:**360–369.

Boissinot, S., Tan, Y., Shyue, S.-K., Schneider, H., Sampaio, I., Neiswanger, K., Hewett-Emmett, D., and Li, W.-H., 1998, Origin and antiquity of the X-linked triallelic color vision systems in New World monkeys, *Proc. Natl. Acad. Sci. USA* **95:**13749–13754.

Deeb, S. S., Lindsey, D. T., Hibiya, Y., Sanocki, E., Winderickx, J., Teller, D. Y., and Motulsky, A. G., 1992, Genotype–phenotype relationships in human red/green color vision defects: molecular and psychophysical studies, *Am. J. Hum. Genet.* **51:**678–700.

Deeb, S. S., Jorgensen, A. L., Battist, L., Iwasaki, L., and Motulsky, A. G., 1994, Sequence divergence of the red and green visual pigments in great apes and humans, *Proc. Natl. Acad. Sci. USA* **91:**7262–7266.

Deegan II, J. F., and Jacobs, G. H., 1996, Spectral sensitivity and photopigment of a nocturnal prosimian, the bushbaby (*Otolemur crassicaudatus*), *Am. J. Primatol.* **40:**55–66.

Dodt, E., 1967, Purkinje shift in the rod eye of the bush baby, *Galago crassicaudatus*, *Vision Res.* **7:**509–517.

Goodman, M., Porter, C. A., Czelusniak, J., Page, S. L., Schneider, H., Shoshani, J., Gunnell, G., and Groves, C. P., 1998, Toward a phylogenetic classification of primates based on DNA evidence complemented by fossil evidence, *Mol. Phyl. Evol.* **9:**585–598.

Hunt, D., Dulai, K. S., Cowing, J. A., Julliot, G., Mollon, J. D., Bowmaker, J. K., Li, W.-H., and Hewett-Emmett, D., 1998, Molecular evolution of trichromacy in primates, *Vision Res.* **38:**3299–3306.

Ibbotson, R., Hunt, D. M., Bowmaker, J. K., and Mollon, J. D., 1992, Sequence divergence and copy number of the middle- and long-wave photopigment genes in Old World monkeys, *Proc. Roy. Soc. Lond. B* **247:**145–154.

Jacobs, G. H., 1984, Within-species variations in visual capacity among squirrel monkeys (*Saimiri sciureus*): Color vision, *Vision Res.* **24:**1267–1277.

Jacobs, G. H., 1993, The distribution and nature of color vision among the mammals, *Biol. Rev.* **68:**413–471.

Jacobs, G. H., 1996, Primate photopigments and primate color vision, *Proc. Natl. Acad. Sci. USA* **93:**577–581.

Jacobs, G. H., and Deegan II, J. F., 1993, Photopigments underlying color vision in ringtail lemurs (*Lemur catta*) and brown lemurs (*Eulemur fulvus*), *Am. J. Primatol.* **30:**243–256.

Jacobs, G. H., and Neitz, J., 1987, Inheritance of color vision in a New World monkey (*Saimiri sciureus*), *Proc. Natl. Acad. Sci. USA* **84:**2545–2549.

Jacobs, G. H., Neitz, J., and Neitz, M., 1993, Genetic basis of polymorphism in the color vision of platyrrhine monkeys, *Vision Res.* **33:**269–274.

Jacobs, G. H., Neitz, M., Deegan, J. F., and Neitz, J., 1996a, Trichromatic colour vision in New World monkeys, *Nature* **382:**156–158.

Jacobs, G. H., Neitz, M., and Neitz, J., 1996b, Mutations in S-cone pigment genes and the absence of colour vision in two species of nocturnal primate, *Proc. R. Soc. Lond. B* **263:**705–710.

Kimura, M., 1980, A simple method for estimating evolutionary rates of base substitutions through comparative studies of nucleotide sequences, *J. Mol. Evol.* **16:**111–120.

Kimura, M., and Ohta, T., 1969, The average number of generations until extinction of an individual mutant gene in a finite population, *Genetics* **63:**701–709.

Klein, J., Takahata, N., and Ayala, F. J., 1993, MHC polymorphism and human origins, *Sci. Am.* **269:**78–83.

Li, W.-H., 1993, Unbiased estimation of the rates of synonymous and nonsynonymous substitution, *J. Mol. Evol.* **36:**96–99.

Li, W.-H., 1997, *Molecular Evolution*, Sinauer, Sunderland, Massachusetts.

Maynard Smith, J., 1992, Analyzing the mosaic structure of genes, *J. Mol. Evol.* **34:**126–129.

Merbs, S. L., and Nathans, J., 1992, Absorption spectra of human cone pigments, *Nature* **356:**433–435.

Merbs, S. L., and Nathans, J., 1993, Role of hydroxyl-bearing amino acids in differentially tuning the absorption spectra of the human red and green cone pigments, *Photochem. Photobiol.* **58:**706–710.

Mervis, R. F., 1974, Evidence of color vision in a diurnal prosimian, *Lemur catta, Anim. Learn. Behav.* **2:**238–240.

Mollon, J. D., Bowmaker, J. K., and Jacobs, G. H., 1984, Variations of colour vision in a New World primate can be explained by a polymorphism of retinal photopigments, *Proc. Roy. Soc. Lond. B* **222:**373–399.

Nathans, J., Thomas, D., and Hogness, D. S., 1986a, Molecular genetics of human color vision: The genes encoding blue, green, and red pigments, *Science* **232:**193–202.

Nathans, J., Piantanida, T. P., Eddy, R. L., Shows, T. B., and Hogness, D. S., 1986b, Molecular genetics of inherited variation in human color vision, *Science* **232:**203–210.

Nei, M., Zhang, J., and Yokoyama, S., 1997, Color vision of ancestral organisms of higher primates, *Mol. Biol. Evol.* **14:**611–618.

Neitz, M., Neitz, J., and Jacobs, G. H., 1991, Spectral tuning of pigments underlying red-green color vision, *Science* **252:**971–974.

Ordy, J. M., and Samorajski, T., 1968, Visual acuity and ERG-CFF in relation to the morphologic organization of the retina among diurnal and nocturnal primates, *Vision Res.* **8:**1205–1225.

Petry, H. M., and Harosi, F. I., 1990, Visual pigments of the tree shrew (*Tupaia belangeri*) and greater, galago (*Galago crassicaudatus*): A microspectrophotometric investigation, *Vision Res.* **30:**839–851.

Reyniers, E., Van Thienen, M.-N., Meire, F., De Boulle, K., Devries, K., Kestelijn, P., and Willems, P. J., 1995, Gene conversion between red and defective green opsin gene in blue cone monochromacy, *Genomics* **29:**323–328.

Saitou, N., and Nei, M., 1987, The neighbor-joining method: A new method for reconstructing phylogenetic trees, *Mol. Biol. Evol.* **4:**406–425.

Schneider, H., Schneider, M. P. C., Sampaio, I., Harada, M. L., Stanhope, M., Czelusniak, J., and Goodman, M., 1993, Molecular phylogeny of the New World monkeys (*Platyrrhini*, primate), *Mol. Phyl. Evol.* **2:**225–242.

Schneider, H., Sampaio, I., Harada, M. L., Barroso, C. M. L., Schneider, M. P. C., Czelusniak, J., and Goodman, M., 1996, Molecular phylogeny of the New World monkeys (*Platyrrhini Primates*) based on two unlinked nuclear genes: IRBP intron 1 and E-globin sequences, *Am. J. Phyl. Anthrop.* **100:**153–179.

Shyue, S.-K., Li, L., Chang, B. H.-J., and Li, W.-H., 1994, Intronic gene conversion in the evolution of human X-linked color vision genes, *Mol. Biol. Evol.* **11:**548–551.

Shyue, S.-K., Hewett-Emmett, D., Sperling, H. G., Hunt, D. M., Bowmaker, J. K., Mollon, J. D., and Li, W.-H., 1995, Adaptive evolution of pigment genes in higher primates, *Science* **269:**1265–1267.

Shyue, S.-K., Boissinot, S., Schneider, H., Sampaio, I., Schneider, M. P., Abee, C. R., William, L., Hewett-Emmett, D., Sperling, H. G., Cowing, J. A., Dulai, K. S., Hunt, D. M., and Li, W.-H., 1998, Molecular genetics of spectral tuning in New World monkey color vision, *J. Mol. Evol.* **46:**697–702.

Spofford, J. B., 1972, A heterotic model for the evolution of duplications, *Brookhaven Symp. Biol.* **23:**121–143.

Sun, H., Macke, K. P., and Nathan, J., 1997, Mechanisms of spectral tuning in mouse green cone pigment, *Proc. Natl. Acad. Sci. USA* **94:**8860–8865.

Takahata, N., and Nei, M., 1990, Allelic geneology under overdominant and frequency-dependent selection and polymorphism of major histocompatibility compex loci, *Genetics* **124:**967–978.

Travis, D. S., Bowmaker, J. K., and Mollon, J. D., 1988, Polymorphism of visual pigment in a callitrichid monkey, *Vision Res.* **28:**481–490.

Wikler, K. C., and Rakie, P., 1990, Distribution of photoreceptor subtypes in the retina of diurnal and nocturnal primates, *J. Neurosci.* **10:**3390–3401.

Winderickx, J., Battisti, L., Motulsky, A. G., and Deeb, S. S., 1992, Selective expression of human X chromosome-linked green opsin genes, *Proc. Natl. Acad. Sci. USA* **89:**9710–9714.

Winderickx, J., Battisti, L., Hibiya, Y., Motulsky, A. G., and Deeb, S. S., 1993, Haplotype diversity in the human red and green opsin genes: Evidence for frequent sequence exchange in exon 3, *Hum. Mol. Genet.* **2:**1413–1421.

Yokoyama, R., and Yokoyama, S., 1990, Convergent evolution of the red- and green-like visual pigment genes in fish, *Astyanax fasciatus*, and human, *Proc. Natl. Acad. Sci. USA* **87:**9315–9318.

Yokoyama, S., and Radlwimmer, F. B., 1998, The "five-sites" rule and the evolution of red and green color vision in mammals, *Mol. Biol. Evol.* **15:**560–567.

Zhao, Z., Hewett-Emmett, D., and Li, W.-H., 1998, Frequent gene conversion between human red and green opsin genes, *J. Mol. Evol.* **46:**494–496.

Zhou, Y.-H., and Li, W.-H., 1996, Gene conversion and natural selection in the evolution of X-linked color vision in higher primates, *Mol. Biol. Evol.* **13:**780–783.

Zhou, Y.-H., Hewett-Emmett, D., Ward, J. P., and Li, W.-H., 1997, Unexpected conservation of the X-linked color vision gene in nocturnal prosimians: evidence from two bush babies, *J. Mol. Evol.* **45:**610–618.

# 9

# The Limits to Knowledge in Conservation Genetics

## The Value of Effective Population Size

LEONARD NUNNEY

## INTRODUCTION

Science is a "way of knowing" (Moore, 1984) that distinguishes itself by developing theories capable of prediction; however, in studies at the interface of evolution and the environment, this task can become formidable. The future direction of evolutionary change is intrinsically unpredictable, because the unit of study (the population) cannot be isolated from changes in its environment. In contrast, the physical sciences and most of biology have been able to achieve the goal of prediction by studying systems that generally can be understood in terms of their internal properties. Thus, a particular type of cell or organism studied today is expected to be much the same when studied by subsequent generations of biologists. However, this expectation is lost when we are asked to consider populations and communities over even moderate periods of time: changes on the time scale of tens of years are commonplace, often dramatic, and often caused by unpredictable events external to the study unit. Short-term changes are primarily numerical, but evolutionary changes, both adaptive (e.g., Reznick *et al.*, 1997) and random (e.g., due to a population bottleneck), can accumulate rapidly. The stochastic models of ecology and population genetics include unpredictable environmental effects, allowing us to make probabilistic pre-

---

LEONARD NUNNEY • Department of Biology, University of California, Riverside, California 92521.

*Evolutionary Biology, Volume 32*, edited by Michael T. Clegg *et al.*
Kluwer Academic / Plenum Publishers, New York, 2000.

dictions that can be quite precise when we consider averages over large numbers of populations, large numbers of genes, or long periods of time. However, we run into a limit to our knowledge when we are asked to make specific predictions about particular systems.

This limit must be confronted by conservation biologists whenever we are asked to plan for the restoration and subsequent long-term survival of a population. We simply cannot know what challenges the population will face in the future. The strategies used to deal with this lack of knowledge can be broken down into two components: demographic and genetic. A demographic plan involves establishing the habitat necessary to maintain the population under the environmental conditions prevailing over the recent past, both biotic and abiotic. This demographic solution, designed to avoid the risk of extinction due to demographic or environmental stochasticity (Shaffer, 1981), is of primary importance to the short-term survival of the population (Lande, 1988). However, contrary to some suggestions (Caro and Laurenson, 1994; Caughley, 1994), the short-term importance of demography does not, *ipso facto*, negate the importance of genetics.

While population genetic factors are often secondary over the short term (but not always; see for example, Saccheri *et al.*, 1998), it does not follow that we can afford to ignore them in creating a conservation plan. It is a truism that all extinction is demographic; however, the effect of genetic changes, or lack of them, on demographic parameters means that they can lead to extinction in a way captured by Gilpin and Soulé (1986) in their vortex metaphor. Random genetic change due to inbreeding can have detrimental effects, as can the lack of an adaptive genetic response to environmental change. Environmental change is intrinsic, even in a world uninfluenced by humans. Climatic change and biotic invasions are features of biological history, often driving extinction and adaptation. Unfortunately, human activities act to increase the magnitude and rate of long-term environmental change and may impose conditions never previously encountered (Lynch, 1996). The ability of a population to adapt to these changes may be paramount for its survival, with the alternative being its extinction.

A possible counterargument to the view that genetic diversity is beneficial is the observation that many invading "weedy" species are relatively (and sometimes very) monomorphic (e.g., dandelions, see King, 1993; Pacific asexual geckos, see Radtkey *et al.*, 1996). In these cases, we observe success with hindsight and we observe it over a relatively short period of time. This is fundamentally different from planning for the future. When an environmental change occurs, there is no way of knowing in advance if a single genotype could be successful, and more importantly, there is no way of knowing what that single genotype might be. Furthermore, even if these questions could be answered, it is likely that the ability of such a genotype to persist through further environmental change would be limited, given the observed failure of asexual lineages to persist through evolutionary time (reviewed by Bell, 1982).

The goal of conservation genetics is to maintain a population that is best able to avoid the detrimental effects of both inbreeding and the inability to adapt. This goal must be achieved using very general criteria, since in general we have no way of knowing the future patterns of natural selection or of predicting how genetic drift may affect specific loci. In this chapter I discuss how the practitioners of conservation genetics attempt to incorporate uncertainty into their plans. I will focus specifically on the problem of the spatial distribution of genetic variance in a conserved population, because a common fate of a threatened population is that it becomes fragmented into several more-or-less isolated units that must be managed as a metapopulation. Spatial subdivision of this type is of particular interest because it alters the relationship between genetic and demographic conservation criteria. In panmictic populations, these two tend to be rather similar; however, the needs of genetic conservation may diverge from those of demographic conservation when spatial patterns are considered (Nunney and Campbell, 1993).

## THE PARAMETERS OF GENETIC PLANNING

The strategies of conservation genetics are based on the assumption that there is a negative correlation between genetic variation and extinction risk. This correlation has two causes. First, very low local levels of genetic variation are generally associated with some form of extreme inbreeding. This, in turn, is associated with inbreeding depression, which lowers fitness, and hence raises the extinction risk (reviewed by Frankham, 1995a). Second, high levels of genetic variability permit a sustained adaptive response to environmental change, lowering extinction risk (Burger and Lynch, 1995). An example of the first is provided by the work of Saccheri *et al.* (1998) on a northern European butterfly. They present strong evidence suggesting that the perseverance of very small populations of this butterfly is determined in large part by inbreeding depression. An example of the second is the work of Grant and Grant (1993) documenting the adaptive response of Darwin's finches to the short-term environmental effects of El Niño. Thus, maintaining genetically variable populations is a high priority for effective conservation. The "ideal" level of genetic variation can be defined operationally as the level typical of the original population prior to anthropogenic threat. However, given the limitations of conservation plans, even this ideal usually must be viewed in the context of a fragmented landscape.

Given the goal of maintaining high levels of genetic variation, some measure must be established to indicate the degree of success achieved or likely to be achieved by a conservation plan. One of the best direct measures of genetic variation is the average heterozygosity of the population, since it can be estimated directly from a relatively small sample. This measure estimates the probability that an individual is heterozygous at a randomly chosen locus, given random mating. Heterozygosity can be used to evaluate current conditions, provided the loci sampled provide a good estimate of the genetic variance present. In the past, problems have arisen with the use of allozyme loci underestimating genetic variability (A. E. Metcalf and L. Nunney, unpublished data). The greater availability of molecular markers lessens this risk; however, the correlation between molecular and quantitative genetic variation can remain weak (Lynch, 1996).

Average heterozygosity has the limitation that it has no predictive value. The current measure of heterozygosity can provide an indication of the immediate evolutionary potential of the population, but it has no necessary relationship to its future value. This is particularly true when the habitat available to the population is changing in the short term, as is often the case when conservation planning becomes a necessity. A better predictive measure is the effective population size ($N_e$). The $N_e$ is the size of an ideal population whose genetic composition is influenced by random processes in the same way as the real population (Wright, 1938). This measure is very important in understanding the genetic structure of small populations, because it allows us to quantify the role of random sampling. In particular, isolated populations lose neutral variation at a rate of $1/(2\,N_e)$ per generation as a result of genetic drift, and they gain variation through mutations that occur at a rate of $2\,N_e u$, where $u$ is the neutral mutation rate. These opposing processes lead to the neutral rate of evolution of $u$ per generation (Kimura, 1968), which is one of the very few features of genetic change in small populations that is independent of $N_e$.

It is not only neutral genetic variation whose dynamics are determined by genetic drift. As populations become smaller, more and more genetic variation becomes "nearly" (or "effectively") neutral. Small selective values that can direct genetic change in large populations no longer do so in sufficiently small ones, at which point the variation becomes nearly neutral. It was shown by the work of Wright (1931) that a rule of thumb defining nearly neutral variation is $N_e s < 1$, where $s$ is the selection differential among the alleles at a locus. One reflection of this result is that, in small populations ($N_e$ of a few hundred or less), quantitative genetic variation acts in a nearly neutral way, since the fitness effect at each locus is small. Under these conditions, the expected genetic variance of a trait is approximately $2\ N_e\sigma^2$, where $\sigma^2$ is the variance added per generation by mutation (Lynch and Hill, 1986). Thus, genetic variation is proportional to $N_e$; only when $N_e$ is larger than ~1,000 does this dependence flatten out (see Lynch, 1996).

This central role of $N_e$ in determining the levels of genetic variation in small populations makes it a very valuable parameter for predicting future changes in the levels of nearly neutral variation. If $N_e$ is very small, then almost all genetic variation becomes nearly neutral, and even highly deleterious alleles can spread through random drift. This possibility, while relevant to conservation in nature, is of particular relevance to captive breeding programs in which the temporary relaxation of natural selection in the breeding environment can make a high proportion of deleterious alleles effectively neutral. The fixation of such deleterious alleles results in permanent inbreeding depression and can result in extinction through a "mutational meltdown" (Lynch *et al.*, 1995). If $N_e$ is larger but still only a few hundred, then the level of quantitative genetic variation is likely to be reduced, with the result that any future adaptive responses will be severely compromised. For these reason, $N_e$ can be expected to be a good management tool for predicting the level of threat facing a population (Mace and Lande, 1991).

## MEASURING EFFECTIVE POPULATION SIZE

The effective population size can be directly measured in two different ways. First, a historical value can be estimated from DNA sequence data using "coalescent" methods, and second, the value prevailing over the recent past can be estimated from gene frequency data using genetic methods. The coalescent approach (see, for example, Fu, 1997; Kuhner *et al.*, 1995) estimates the parameter $\theta$ (= $4\ N_eu$), which, combined with plausible estimates of $u$, allows us to estimate $N_e$. The method projects back into the past to provide the estimate of $\theta$ that best describes the data, given of course the assumptions of the estimation model. Genetic methods attempt to measure the effect of genetic drift over very few generations, from which $N_e$ can be estimated. The most commonly used approach is the temporal method (Waples, 1989), in which drift is estimated from the changes in gene frequency measured over several generations (e.g., Jorde and Ryman, 1996). Two other genetic methods have been suggested that have the advantage of requiring only a single sampling period. One is based on the measurement of randomly generated linkage disequilibrium (see Bartley *et al.*, 1992) and the other is based on the excess of heterozygotes observed in small populations (Pudovkin *et al.*, 1996). Unfortunately, both of these single-sample techniques are dependent on the assumption of random mating. For example, the heterozygote excess is easily erased by any spatial patterning of the gene frequency (due to the Wahlund effect).

An alternative approach to the estimation of $N_e$ is to predict its value based on known properties of the population. This is the ecological method (Nunney and Elam, 1994), and it is based on theoretical links between $N_e$ and demographic and behavioral factors. For example, Wright (1938) established that, when the generations are nonoverlapping, $N_e$ is proportional to $N$, the adult population size, but is reduced by increasing variance in individual reproductive success ($V$):

$$N_e = \frac{4N}{(2+V)} \tag{1}$$

(given $N >> 1$). Note that in a stable population, $V = 2$ reflects Poisson variance (= mean reproductive success); hence, $N_e = N$.

Equation (1) can be generalized to overlapping generations (Hill, 1972; Nunney, 1993), and when these equations are parameterized using data from animal populations, it is predicted that the ratio of $N_e/N$ will typically be in the range 0.25–0.75, provided the effects of a fluctuating adult population can be factored out by using an appropriate measure of $N$ (Nunney and Elam, 1994; Nunney, 1996). When the generations are nonoverlapping, this can be approximated using the harmonic mean of $N$, since Wright (1938) noted that $N_e$ could be averaged over multiple generations in this way. No equivalent measure has been shown to apply when the generations overlap. Field data indicate that measuring $N$ by its arithmetic mean has the effect of lowering the ratio $N_e/N$ from the expected median of approximately 0.5 to about 0.1–0.2 (Nunney and Campbell, 1993; Frankham, 1995b; Vucetich *et al.*, 1997). Thus, the ecological approach, to be predictive, must include an analysis of the population dynamic patterns of the species.

From the perspective of conservation, the advantage of the ecological approach is that, unlike the other two, it can be used as a tool for predicting $N_e$ in the future. Estimates of $N_e$ based on coalescent or genetic methods provide no information concerning why $N_e$ has a particular value, and hence provide no information on what would happen under some modified set of conditions prescribed by a conservation plan. The ideal study, often difficult to achieve, is to combine both genetic and ecological estimates (Begon *et al.*, 1980; Husband and Barrett, 1992; see also Nunney, 1995), since this allows a validation of the indirect ecological approach with the direct genetic approach.

## POPULATION FRAGMENTATION—THE ISLAND MODEL

A feature of most conservation plans is the subdivision of an originally continuous habitat into a series of isolated reserves. To evaluate the population genetic consequences of such fragmentation, we must consider two levels: all the reserves considered together (the metapopulation level) and each reserve considered individually. We need to understand both the rate of loss of existing genetic variation from the whole metapopulation and the projected distribution of genetic variation across the reserves. These two issues are related, and together they can have important consequences for long-term population viability.

Wright (1943) showed that when a metapopulation of size $N$ is divided up into a series of identical semi-isolated islands, each of which contributes equally to the pool of interisland migrants, then the effective size of the whole metapopulation is:

$$N_e = \frac{N}{(1 - F_{ST})} \tag{2}$$

where $F_{ST}$ is Wright's hierarchical inbreeding coefficient measuring interisland genetic differentiation. The expected equilibrium value of $F_{ST}$ in the island model is $1/(1 + 4\ n_e m)$, where $n_e$ is the effective size of each island considered in isolation and $m$ is the migration rate (and assuming $m \ll 1$). Thus, for low migration rates:

$$N_e = [1 + 1/(4\ n_e m)]N \tag{3}$$

From Eqs. (2) and (3), it can be seen that as $m \rightarrow 0$ (and hence $F_{ST} \rightarrow 1$), the effective size of the metapopulation becomes infinite. Given the benefits of a large $N_e$ outlined above, it would seem that the island scenario is in many respects ideal; however, there are important problems with this conclusion once interisland migration disappears.

To understand these problems, we must first consider the genetic benefits of the island structure. Subdivision slows the rate of loss of genetic variation by acting to isolate the effects of drift within the islands. For example, in the extreme case of complete isolation, the effects of drift are limited to within islands. Thus, an allele that spreads to fixation in one island can spread no further, i.e., $N_e$ is infinite. This same isolation of the effects of drift also leads to the expectation that the genetic variance for a quantitative trait maintained under a drift–mutation balance within a metapopulation is increased by subdivision (see Lande, 1995a). Furthermore, provided that the islands are linked by migration, then the within-island variance is independent of the migration rate (Lynch, 1994). Barton and Whitlock (1997) illustrated a similar point with a simple model of stabilizing selection.

If the islands of a metapopulation become completely isolated from each other, then the beneficial effects either disappear or cease to be beneficial. At single loci, homozygosity is expected to be the norm, unless $n_e u > 1$ (Wright, 1931), where $u$ is the per locus mutation rate. In the extreme case, with $n_e$ small enough that we can ignore mutation, each island will contain only a single multilocus genotype. Across quantitative trait loci, there is a similar effect as the within-island level of genetic variance declines to an equilibrium determined entirely by island size (approximately 2 $n_e \sigma^2$; see above). These effects create two problems. First, the potentially dramatic long-term drop in within-island variability has the effect that the isolated island populations may be unable to evolve in response to changes in the environment. Second, within islands, some disadvantageous alleles (with fitness $1 - s$) will be effectively neutral (i.e., those satisfy $n_e s < 1$). The fixation of a proportion of such alleles results in inbreeding depression.

These effects suggest that $N_e$, as defined by Eq. (2), is not always a useful tool for conservation planning. An operational measure of effective size that reflects processes at both the metapopulation and reserve levels is $N_{e,min}$, defined as the smaller of $N_e$ (the effective size of the metapopulation) and $N_{e,sum}$ (the sum of the $n_e$ over all reserves). The logic of this measure is that when $N_e > N_{e,sum}$, then the reserves are acting as more-or-less independent evolutionary units and $N_{e,sum}$ reflects this independence, and when $N_{e,sum} > N_e$, then the reserves are integrated into a single evolutionary unit of size $N_e$. Thus, $N_{e,min}$ allows us to retain effective size as a predictor of the genetic quality of the metapopulation.

Given an island model, $N_e$ is always greater than $N_{e,sum}$ (except under very unusual circumstances). Integration of the islands into a single evolutionary unit is precluded by the assumption that all islands are equally productive. This assumption is reasonable when migration is zero (i.e., the productivity of each island exactly replaces itself), but any nonzero migration rate will result in some variation among islands, and the island model no longer applies.

## POPULATION FRAGMENTATION—THE INTERDEMIC MODEL

Relaxing the assumption that all islands produce exactly the same number of migrants changes the model to one of interdemic selection. This can be modeled with migration made up of two components. The first component is fixed and invariant across islands (demes), just as in the island model, and the second is proportional to the productivity of each island. Thus, the migration rate can never be zero, since even if the fixed component is zero, productivity variation leads to some dispersal. For example, under the assumption of Poisson variation in female reproductive success, the variance in productivity is fixed at $1/n$, where $n$ is the size of each deme.

Following Wright's (1943) derivation of Eq. (2), but allowing the random Poisson differences in female fecundity to result in differences in island productivity, results in a new relationship (Nunney, 1999):

$$N_e = \frac{N}{(1 + F_{ST})} \tag{4}$$

Under the interdemic model, the effective size of a metapopulation is always reduced by increasing $F_{ST}$, the precise opposite of the island model. The effect is driven by the action of interdemic genetic drift, which is precluded in the island model.

As noted above, this interdemic scenario cannot describe a situation where there is no migration (and hence no productivity variance) among demes, nor where local population regulation removes all productivity differences among populations. However, even small differences in migrant production can have a large effect. For example, under otherwise ideal conditions, an interdemic variance of half that expected under Poisson sampling will lower the effective size of the population from that predicted by Eq. (2) down to $N_e = N$.

From a conservation perspective, one very important potential source of productivity variation is extinction. If an island subpopulation is lost, then its productivity is zero and the productivity of the islands used for recolonization is increased. The effect of extinction–recolonization cycles on $N_e$ can be calculated using a result of Whitlock and Barton (1997) that incorporates the variance in island productivity ($V_s$):

$$N_e = \frac{N}{(1/2 + V_i)(1 - F_{ST}) + 2\,nF_{ST}V_s\,k/(k-1)} \tag{5}$$

where $V_i$ is the standardized variance in reproductive success among individuals within islands, and $n$ and $k$ are the size and number of islands, respectively. The original relationship has been modified to include $V_i$ (see Nunney, 1999).

If the variance in island productivity is due to extinction–recolonization, with $e$ as the probability of extinction and $m$ the migration rate, then (approximately):

$$N_e = \frac{k}{4(m+e)F_{ST}} \tag{6}$$

(Whitlock and Barton, 1997). Note in particular that in Eq. (6), $N_e$ is proportional not to $N$, the total population size, but to $k$, the number of islands. As a result, extinction–recolonization can result in very low values of $N_e$, a result suggested by the very rapid loss of genetic variation observed in the simulations of Gilpin (1991) and Hedrick and Gilpin (1996).

## MOVEMENT AMONG RESERVES

Many conserved populations are subdivided among a number of island reserves. These reserves may be completely isolated from each other, or they may be linked either by habitat corridors or through the ability of the organism to disperse across unfavorable habitat. Each of these possibilities raises questions over the best strategy for managing interreserve movement. The most appropriate strategy will depend both on the options available for interisland movement and on knowledge of the prefragmentation pattern of genetic variation. A fair estimate of this prefragmentation pattern often can be deduced from a knowledge of the original distribution of the species, in combination with its dispersal biology. The absence of historical genetic differentiation at neutral loci over the region would suggest that any local adaptation and/or coadaptation was stable under conditions of gene flow and would favor maintaining relatively high levels of gene flow. However, in a highly genetically differentiated population, much lower gene flow might be appropriate.

In most cases, the anthropogenic fragmentation of metapopulations will result in lower levels of gene flow than prevailed historically. This has the potentially beneficial effect of promoting increased local adaptation. But increased local adaptation does not negate the value of maintaining sufficient gene flow to maintain a diverse local gene pool. Without movement, drift and perhaps selection will drive differentiation, so that eventually most of the genetic variance will be distributed among the reserves, whereas individual reserve populations will be genetically homogeneous. This distribution of genetic variance can create severe problems since the bulk of the variance is unusable for adaptive change to a new environmental challenge. Any change will rely on the small amount of variation left within each reserve, since recombination across reserve gene pools is not possible.

The management of migration can be complicated by several factors. Wright (1931) showed that movement of one to two "effective" individuals per generation is sufficient to prevent significant genetic differentiation at a single locus due to drift, and this translates into successfully moving perhaps two to four adults per generation. If the progeny of the immigrants have lowered fitness (because they are not locally adapted), this could lead to the near absence of gene flow in spite of a program of regular translocations. Dispersal routes between reserves, either due to corridors or the natural dispersal ability of the species, have the great advantage of precluding the need for artificially moving individuals. However, if some reserves contribute more migrants than others, either predictably (because some reserves have superior habitat) or randomly [as assumed for Eq. (5)], then $N_e$ will be lowered. The effect may be impossible to avoid, but the possibility of a lowered overall $N_e$ should be taken into account when the reserve system is established.

Population extinction is always a possibility, and from a demographic perspective the possibility generally favors some subdivision. The chance of a single, large population going extinct through some localized environmental effect is generally greater than the chance of a number of small subpopulations all going extinct simultaneously (see Lande, 1993). Thus, multiple reserves can be seen as a form of insurance against complete loss. However, in considering the degree of subdivision, genetic considerations become important if local extinction is a nontrivial risk. As shown by Eq. (6), extinction–recolonization is a very efficient mechanism for removing genetic variation from a metapopulation: Each time a subpopulation goes extinct, substantial unique genetic variation may disappear. While the number of individuals lost can often be made up very quickly through recolonization, the genetic loss is likely to be long term. Thus, although demographic considerations are not in conflict with the practical expedient of allowing a failing subpopulation to go extinct prior to recolonization, genetic considerations argue strongly that extinction of local populations should be avoided whenever possible and favor active intervention to prevent complete extinction and consequent loss of valuable genetic variation.

If a reserve population does go extinct, how should it be recolonized? Using a small number of individuals from a single reserve is the least satisfactory method. This method introduces a limited sample of the genetic variance from a single reserve and amplifies it. The result is that extinction–recolonization leads to a rapid deterioration in genetic variation of the metapopulation. If all other things are equal, the best method is to introduce a sample from all extant subpopulations. This process acts to create a new reservoir of genetic variation each time an extinction occurs and slows its erosion. Balanced against this is the possible importance of outbreeding depression (see Waser, 1993). However, recolonization, if possible, always should be made with individuals from more than one reserve and in sufficient numbers to avoid a substantial founder effect.

It has been argued that under some circumstances the movement of individuals between reserves should be avoided because of the possibility of disease transmission (see, for example, Hess, 1994). However, there are two compelling genetic arguments against this strategy: first, it acts to lessen the likelihood of resistance evolving; second, it probably does little to prevent the spread of the diseases expected in endangered species. Perhaps more than any other component of the environment, diseases drive genetic change in their hosts. The most polymorphic vertebrate genes, the genes of the major histocompatibility complex, are involved in disease resistance. In fact, this involvement prompted Hughes (1991) to suggest that maintaining this variation should be a primary goal of captive breeding programs (but see Gilpin and Wills, 1991; Miller and Hedrick, 1991). The likelihood of evolving resistance to a disease will increase with the levels of genetic variation within a reserve population, but preventing movement among reserves will act to lower the available genetic variation. Furthermore, isolating the reserves of a conserved species will probably do little to prevent disease spread. It is expected that many diseases of rare organisms will come from the common organisms that surround them (see Daszak *et al.*, 2000). It is certainly notable that two of the most cited examples, rinderpest and canine distemper, are generally introduced as a direct or indirect result of human activity (see Hess, 1994). Rather than try to isolate a conserved species from disease, which is generally impossible, we should do everything we can to maximize the likelihood that it can adapt to any new threat.

## CONCLUSION

The use of effective population size as an important tool of conservation genetics is an implicit recognition that we cannot know the future challenges that will confront a population. By maximizing $N_e$, we are planning for the unknown by attempting to maximize the genetic variation available for an adaptive response. Some attempts have been made to suggest a minimum value of $N_e$ likely to achieve this goal. Franklin (1980) suggested $N_e = 500$ as an acceptable minimum, whereas Lande (1995b) pointed out that a target of $N_e = 5{,}000$ is more appropriate, based on the expected additive variance for quantitative traits in very large populations. A more context-specific ideal, outlined earlier, has as its goal the maintenance of a level of genetic variation close to that of the original population prior to anthropogenic threat. It seems probable that this goal may often be realized with $N_e$ between 500 and 5,000. An important and all-too-often forgotten aspect of these rough guidelines, is that $N_e$ is not the same as $N$. If a minimum $N_e$ of 500 is the conservation goal, then the carrying capacity of the habitat would almost certainly have to be several thousand for there to be much chance of maintaining $N_e = 500$ over the long term (Nunney and Campbell, 1993).

There is no "right" answer (except infinity) to the problem of defining the minimum long-term $N_e$. A serious gap in our knowledge concerns just how species-specific such minima may be. As yet we do not know if there are some aspects of biology or ecology that predict which species can survive a relatively low level of genetic variability. The low level of genetic variation in the cheetah (Menotti-Raymond and O'Brien, 1993) is a case in point. As a species, it may be living on borrowed time, or there may be some feature of its biology that makes it more able to survive with limited genetic variation. Answers to these questions will allow us to tailor conservation plans more precisely.

Spatial subdivision of a population complicates the interpretation of $N_e$. At the extreme, a population divided into complete isolates has an infinite effective size and yet cannot exploit its global genetic variance, since variation among the isolates cannot be recombined. This issue has practical importance. There are a growing number of habitat conservation plans in which, without direct intervention, movement among reserves is precluded. These groups of reserves need to be managed as a complete unit. First, movement plans need to be developed to prevent inbreeding and prevent the genetic divergence of the reserve populations. Second, it is important that the genetic dangers of extinction–recolonization cycles are incorporated into the management plans for groups of reserves. If isolated reserves are managed individually, then the long-term danger is that it will be necessary to independently manage a number of small vulnerable populations that require high levels of intervention instead of managing a single large resilient metapopulation.

## REFERENCES


Bartley, D., Bagley, M., Gall, G., and Bentley, B., 1992, Use of linkage disequilibrium data to estimate effective size of hatchery and natural fish populations, *Cons. Biol.* **6:**365–375.

Barton, N. H., and Whitlock, M. C., 1997, The evolution of metapopulations, in: *Metapopulation Biology* (I. Hanski and M. E. Gilpin, eds.), pp. 183–210, Academic Press, New York.

Begon, M., Krimbas, C. B., and Loukas, M., 1980, The genetics of *Drosophila subobscura* populations. XV. Effective size of a natural population estimated by three independent methods, *Heredity* **45:**335–350.

Bell, G., 1982, *The Masterpiece of Nature: The Evolution and Genetics of Sexuality*, University of California Press, Berkeley.

Burger, R., and Lynch, M., 1995, Evolution and extinction in a changing environment: A quantitative genetic analysis, *Evolution* **49:**151–163.

Caro, T. M., and Laurenson, M. K., 1994, Ecological and genetic factors in conservation: A cautionary tale, *Science* **263:**485–486.

Caughley, G., 1994, Directions in conservation biology, *J. Anim. Ecol.* **63:**215–244.

Daszak, P., Cunningham, A. A., and Hyatt, A. D., 2000, Emerging infectious diseases of wildlife—Threats to biodiversity and human health, *Science* **287:**443–449.

Frankham, R., 1995a, Conservation genetics, *Annu. Rev. Genet.* **29:**305–327.

Frankham, R., 1995b, Effective population size—Adult population size ratios in wildlife: A review, *Genet. Res.* **66:**95–107.

Franklin, I. R., 1980, Evolutionary changes in small populations, in: *Conservation Biology: An Evolutionary–Ecological Perspective* (M. E. Soulé and B. A. Wilcox, eds.), pp. 135–149, Sinauer Associates, Sunderland, Massachusetts.

Fu, Y-X., 1997, Coalescent theory for a partially selfing population, *Genetics* **146:**1489–1499.

Gilpin, M., 1991, The genetic effective size of a metapopulation, *Biol. J. Linn. Soc.* **42:**165–175.

Gilpin, M. E., and Soulé, M. E., 1986, Minimum viable populations: Processes of species extinctions, in: *Conservation Biology: The Science of Scarcity and Diversity* (M. E. Soulé, ed.), pp. 19–34, Sinauer Associates, Sunderland, Massachusetts.

Gilpin, M., and Wills, C., 1991, MHC and captive breeding: A rebuttal, *Cons. Biol.* **5:**554–555.

Grant, B. R., and Grant, P. R., 1993, Evolution of Darwin's finches caused by a rare climatic event, *Proc. Roy. Soc. Lond. B* **251:**111–117.

Hedrick, P. W., and Gilpin, M. E., 1996, Genetic effective size of a metapopulation, in: *Metapopulation Dynamics: Ecology, Genetics, and Evolution* (I. A. Hanski and M. E. Gilpin, eds.), pp. 165–181, Academic Press, New York.

Hess, G. R., 1994, Conservation corridors and contagious disease: A cautionary note, *Cons. Biol.* **8:**256–262.

Hill, W. G., 1972, Effective size of populations with overlapping generations, *Theor. Pop. Biol.* **3:**278–289.

Hughes, A. L., 1991, MHC polymorphism and the design of captive breeding programs, *Cons. Biol.* **5:**249–251.

Husband, B. C., and Barrett, C. H., 1992, Effective population size and genetic drift in tristylous *Eichornia paniculata* (Pontederiaceae), *Evolution* **46:**1875–1890.

Jorde, P. E., and Ryman, N., 1996, Demographic genetics of brown trout (*Salmo trutta*) and estimation of effective population size from temporal change of allele frequencies, *Genetics* **143:**1369–1381.

Kimura, M., 1968, Evolutionary rate at the molecular level, *Nature* **217:**624–626.

King, L. M., 1993, Origins of genotypic variation in north American dandelions inferred from ribosomal DNA and chloroplast DNA restriction enzyme analysis, *Evolution* **47:**136–151.

Kuhner, M. K., Yamato, J., and Felsenstein, J., 1995, Estimating effective population size and mutation rate from sequence data using Metropolis–Hastings sampling, *Genetics* **140:**1421–1430.

Lande, R., 1988, Genetics and demography in biological conservation, *Science* **241:**1455–1460.

Lande, R., 1993, Risks of population extinction from demographic and environmental stochasticity, and random catastrophes, *Am. Natur.* **142:**911–927.

Lande, R., 1995a, Breeding plans for small populations based on the dynamics of quantitative trait variance, in: *Population Management for Survival and Recovery* (J. D. Ballou, M. Gilpin, and T. J. Foose, eds.), pp. 318–340, Columbia University Press, New York.

Lande, R., 1995b, Mutation and conservation, *Cons. Biol.* **9:**782–791.

Lynch, M., 1994, Neutral models of phenotypic evolution, in: *Ecological Genetics* (L. A. Real, ed.), pp. 86–108, Princeton University Press, Princeton, New Jersey.

Lynch, M., 1996, A quantitative-genetic perspective on conservation issues, in: *Conservation Genetics: Case Histories from Nature* (J. C. Avise, and J. L. Hamrick, eds.), pp. 471–501, Chapman and Hall, New York.

Lynch, M., and Hill, W. G., 1986, Phenotypic evolution by neutral mutation, *Evolution* **40:**915–935.

Lynch, M., Conery, J., and Burger, R., 1995, Mutation accumulation and the extinction of small populations, *Am. Natur.* **146:**489–518.

Mace, G. M., and Lande, R., 1991, Assessing extinction threats: Towards a reevaluation of IUCN threatened species categories, *Cons. Biol.* **5:**148–157.

Menotti-Raymond, M., and O'Brien, S. J., 1993, Dating the genetic bottleneck of the African cheetah, *Proc. Natl. Acad. Sci. USA* **90:**3172–3176.

Miller, P. S., and Hedrick, P. W., 1991, MHC polymorphism and the design of captive breeding programs: Simple solutions are not the answer, *Cons. Biol.* **5:**556–558.

Moore, J. A., 1984, Science as a way of knowing—Evolutionary biology, *Am. Zool.* **24:**467–534.

Nunney, L., 1993, The influence of mating system and overlapping generations on effective population size, *Evolution* **47:**1329–1341.

Nunney, L., 1995, Measuring the ratio of effective population size to adult numbers using genetic and ecological data, *Evolution* **49:**389–392.

Nunney, L., 1996, The influence of variation in female fecundity on effective population size, *Biol. J. Linn. Soc.* **59:**411–425.

Nunney, L., 1999, The effective size of a hierarchically structured population, *Evolution* **53:**1–10.

Nunney, L., and Campbell, K. A., 1993, Assessing minimum viable population size: Demography meets population genetics, *Trends Ecol. Evol.* **8:**234–239.

Nunney, L., and Elam, D. R., 1994, Estimating the effective size of conserved populations, *Cons. Biol.* **8:**175–184.

Pudovkin, A. I., Zaykin, D. V., and Hedgecock, D., 1996, On the potential for estimating the effective number of breeders from heterozygote-excess in progeny, *Genetics* **144:**383–387.

Radtkey, R. R., Becker, B., Miller, R. D., Riblet, R., and Case, T. J., 1996, Variation and evolution of Class I MHC in sexual and parthenogenetic geckos, *Proc. Roy. Soc. Lond. B* **263:**1023–1032.

Reznick, D. N., Shaw, F. H., Rodd, F. H., and Shaw, R. G., 1997, Evaluation of the rate of evolution in natural populations of guppies (*Poecilia reticulata*), *Science* **275:**1934–1937.

Saccheri, I., Kuussaari, M., Kankare, M., Vikman, P., Fortelius, W., and Hanski, I., 1998, Inbreeding and extinction in a butterfly metapopulation, *Nature* **392:**491–494.

Shaffer, M. L., 1981, Minimum population sizes for species conservation, *Bioscience* **31:**131–134.

Vucetich, J. A., Waite, T. A., and Nunney, L., 1997, Fluctuating population size and the ratio of effective to census population size ($N_e/N$), *Evolution* **51:**2017–2021.

Waples, R. S., 1989, A generalized approach for estimating effective population size from temporal changes in allele frequency, *Genetics* **121:**379–391.

Waser, N. M., 1993, Population structure, optimal outbreeding, and assortative mating in angiosperms, in: *The Natural History of Inbreeding and Outbreeding: Theoretical and Empirical Perspectives* (N. W. Thornhill, ed.), pp. 173–199, University of Chicago Press, Chicago.

Whitlock, M. C., and Barton, N. H., 1997, The effective size of a subdivided population, *Genetics* **146:**427–441.

Wright, S., 1931, Evolution in Mendelian populations, *Genetics* **16:**97–159.

Wright, S., 1938, Size of population and breeding structure in relation to evolution, *Science* **87:**430–431.

Wright, S., 1943, Isolation by distance, *Genetics* **28:**114–138.

# 10

# What Is the Structure of Human Populations?

BRUCE S. WEIR

## INTRODUCTION

The study of evolution is limited by its essentially stochastic nature. Even if we understand all the forces affecting the transmission of genes from one generation to the next, we cannot predict the future genetic constitution of a population and we cannot reconstruct all aspects of the past. The phrase "cannot predict" is to be taken in the same sense as when we decline to predict whether a toss of a fair coin will result in a head or a tail. We may be content to assign a probability of one half to each outcome, but the inherent uncertainty in the tossing process is widely acknowledged and even serves as the basis for deciding elections in the case of equal numbers of votes for each of two candidates. The possibility of studying the physics of coin tossing to such a degree that the outcome can be predicted with certainty is rejected, and biological processes such as meiosis are very much more complicated. Our theories of evolution are designed to apply on the average or to give probabilities to sets of outcomes. Our data, however, are from specific outcomes. Our understanding of evolution must proceed from the single replicate of the process. We can no more construct a complete theory of a stochastic process from specific data than we can predict a specific outcome with certainty from our theory.

In spite of this, we have achieved a great deal of understanding of the evolutionary process, and we do make valuable predictions in plant and

BRUCE S. WEIR • Program in Statistical Genetics, Department of Statistics, North Carolina State University, Raleigh, North Carolina 27695-8203.

*Evolutionary Biology, Volume 32*, edited by Michael T. Clegg *et al.*
Kluwer Academic / Plenum Publishers, New York, 2000.

animal breeding and in conservation biology. I propose to discuss this apparent contradiction by making reference to two areas in which evolutionary biology provides the framework for inference and in which the structure of human populations is an important factor: These are the areas of human disease gene mapping by marker associations and the interpretation of matching DNA profiles for human identification. Both areas are hampered by a lack of knowledge of population structure.

## HUMAN DISEASE GENE MAPPING

Locating the gene(s) responsible for a human disease is a necessary step in the path to characterizing the gene(s) and eventually curing the disease. Linkage studies, in which the transmissions of disease and marker alleles are followed down disease pedigrees (Ott, 1991), have been very successful in locating many genes to within about 1 centiMorgan (cM) of a marker. Although it is the physical position that is needed, linkage studies estimate map distance (actually recombination fraction) between putative disease genes and marker locus. Any distances smaller than 1 cM result in there being too few recombination events to allow such events to be detected in pedigrees of practical size. Estimates of such small values is not possible. The hope has been that smaller recombination fractions can be estimated from associations between disease status and marker type.

Initial efforts invoked population genetic theory for populations in a state of evolutionary equilibrium. For such populations, the amount of linkage disequilibrium between loci is expected to be zero, but the expected value of the squared correlation, $r^2$, of allele frequencies has a nonzero value. A common result equates this to a function of the compound parameter 4 $Nc$:

$$\varepsilon(r^2) = \frac{1}{1+4Nc}$$

Here $N$ is the population size and $c$ the recombination fraction.

There has been some tension between experimentalists, who took this equation as justification for estimating 4 $Nc$ from

$$\widehat{4Nc} = \frac{1}{r^2} - 1$$

and theoreticians, who pointed out that this was equivalent to equating a single observation to the mean of the distribution from which it was drawn (Weir and Hill, 1986).

More recent activity in this area has abandoned the assumption of evolutionary equilibrium and instead has assumed that the observation of linkage disequilibrium between disease and marker is indicative of a young disease: There has been at most a few hundred generations since the mutation to the disease allele. The observed association is a consequence of there being too little time since the ancestral disease mutation for recombination to have broken up the initial complete disequilibrium. Once again, there are simple results available from population genetic theory. In the simplest model, if the disease mutation in the present population is $G$ generations old, then current chromosomes carry both the disease allele and the marker allele initially associated with the mutant only if there has been no recombination over $G$ generations:

$$\varepsilon(p_{excess}) = (1-c)^G \approx 1 - cG$$

where $p_{excess}$ is the (relative) excess of the frequency of marker allele on disease versus normal chromosomes. There is a temptation to estimate $Gc$ from

$$\widehat{Gc} = 1 - p_{excess}$$

Although the young disease model is likely to be more appropriate than an evolutionary equilibrium model, there is still the problem that a single realization of the evolutionary process provides a single value of a quantity such as $r^2$ or $p_{excess}$, whereas population genetic theory provides expected values.

One method of proceeding is to use a computer to simulate the evolutionary process under the assumed model (Hill and Weir, 1994, for an equilibrium model; Kaplan *et al.*, 1995, for a young disease model) and use the resulting distribution to formulate a likelihood from the observed data. If the simulation process is repeated for many values of $c$ (via 4 $Nc$ or $Gc$), then the value of $c$ that maximizes the likelihood provides a point estimate, and the likelihood curve leads to confidence (or support) intervals around this estimate. These simulations recognize the stochastic nature of evolution, and take the place either of actual replication or of a theory that describes the variation in two-locus frequencies over evolutionary replicates.

One alternative would be to seek empirical replication of the evolutionary process. The use of different populations, or different loci, attempts to overcome the inherent limitation of a single sample. Of course, questions about associations of disease genes to specific markers are unlikely to be addressed by looking at different markers, and results for a single marker could well differ between populations in which the disease had different ages or marker backgrounds.

The alternative of having a theory in place that will give the distribution of the quantities of interest over evolutionary replicates is not available. Although statistics provides a sound theory for variation among repeated samples from the same population, there is no equivalent theory for variation among repeated populations. In the language of Weir (1996), multinomial theory often can be used for variation in allele frequencies due to statistical sampling, but genetic sampling requires a very much more complex treatment. The most successful theory is that of Wright (1951), who showed the applicability of a Dirichlet distribution, but even that required an evolutionary equilibrium that clearly does not hold for human populations.

Even a complete theory, or exhaustive simulations, can allow only estimates and their uncertainties to be obtained from data. The exact evolutionary history cannot be reconstructed, so that the actual recombination value cannot be determined by population association methods.

## HUMAN IDENTIFICATION

When the same DNA profile is found in a biological sample and in a person thought to have provided that sample, the appropriate way to express the strength of this evidence is to compare the probabilities of the evidence under alternative propositions (Evett and Weir, 1998). One of these propositions may be that the person did provide the sample, and then the evidence is certain. The other proposition may be very specific, or may simply be that some other person was the source of the sample. In that case, the probability of the evidence involves the probability of any other untyped person having the profile in question, *given that the named person has the profile*, and this has to be considered for every other untyped person. Clearly, family members of the named person have a higher chance than non–family members of having the profile. Less clearly, but possibly of more importance, is the fact that people with a closer evolutionary relationship to the named person are more likely than others to have the profile.

This argument has led to the use of conditional genotype probabilities of the form, for $AA$ homozygotes say,

$$\Pr(AA|AA) = \left(p_A^*\right)^2$$

where $p_A^*$ is the frequency of allele $A$ in the subpopulation to which both the named person and the untyped other person both belong. There is an implicit assumption of random mating within the subpopulation, and the answer is conditioned on $p_A^*$. If data are not available from the subpopulation, then the conditional probability is

$$\Pr(AA|AA) = \frac{[2\theta + p_A(1-\theta)][3\theta + p_A(1-\theta)]}{(1+\theta)(1+2\theta)}$$

(Balding and Nichols, 1994), where $p_A$ is the populationwide frequency for allele $A$ and $\theta$ is the coancestry coefficient within the subpopulation to which the two people both belong. This conditional probability formulation recognizes that the variance of allele frequencies over subpopulations is given by $\theta p_A(1 - p_A)$.

The first problem is that $\theta$ is unknown. It can be estimated only when information on variation among subpopulations is known, and it is the lack of that information that led to the need for conditional probabilities involving $\theta$ in the first place. The problem is both conceptual and practical. At the practical level, it may not be feasible to collect data from all possible subpopulations. Conceptually, it is not clear how these subpopulations are to be defined.

It has been less-well appreciated that the allele frequencies $p_A$ are also unknown (I. S. Painter and B. S. Weir, unpublished observations). The conditional probability formulation requires that $\theta$ refers to allelic relationships relative to a reference population, and it is this reference population to which $p_A$ refers. It is not appropriate to use observed frequencies from the current total population, since that would ignore the variation among replicate populations descending from the same reference population. This point is explored further in the next section.

## Hierarchical Model of Human Evolution

A grossly simplified model of the evolution of modern humans envisages a series of bifurcations, starting with the African–non-African split. Such a model suggests the idealized representation of Fig. 1. In that figure, individuals within the same subpopulation have the longest shared ancestry, so we expect

$$\theta_S \geq \theta_P \geq \theta_R \geq \theta_A$$

where $\theta_S$, $\theta_P$, $\theta_R$, $\theta_A$, respectively, indicate the coancestry of two individuals in the same subpopulation, individuals in different subpopulations of the

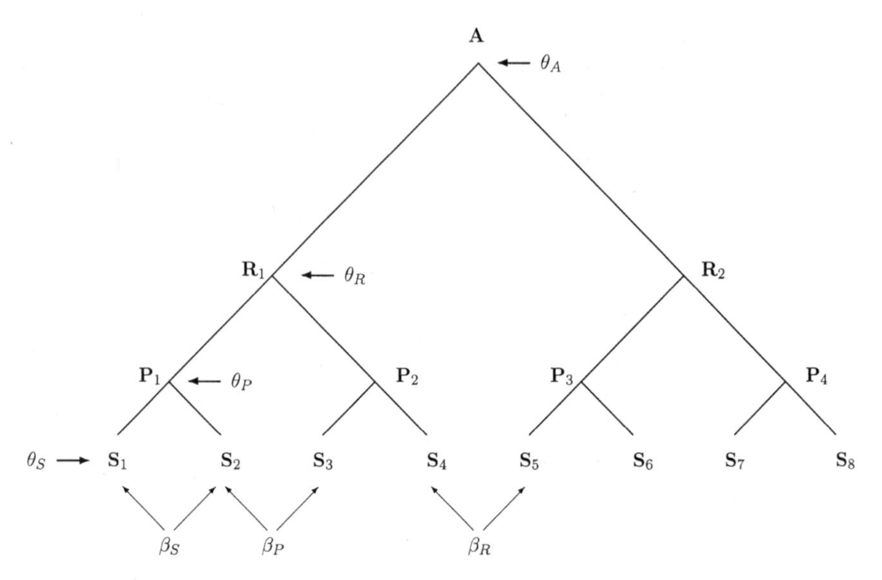


FIG. 1. Idealized evolutionary history (A, ancestral; R, race; P, population; S, subpopulation).

same population, individuals in different subpopulations in different populations of the same race, and individuals in different subpopulations of different races.

If the same value of $\theta$ is assumed to hold for all subdivisions at the same level in the hierarchy, then it is straightforward to estimate the $F$ statistics. Suppose the traditional parameter $F_{ST}$ is estimated from a pair of subpopulations (say $S_1, S_2$) within the same population. This will provide an estimate of

$$\beta_S = \frac{\theta_S - \theta_P}{1 - \theta_P}$$

(Cockerham, 1973). Likewise, data from two subpopulations (say $S_1$, $S_3$) within different populations of the same race will produce an $F_{ST}$ estimate of

$$\beta_P = \frac{\theta_S - \theta_R}{1 - \theta_R}$$

whereas data from two subpopulations (say $S_1$, $S_5$) within different races will give an estimate of

$$\beta_R = \frac{\theta_S - \theta_A}{1 - \theta_A}$$

For a wide variety of evolutionary models, these estimates will have the relationship

$$\beta_S \leq \beta_P \leq \beta_R$$

and can serve as measures of distance between subpopulations at the three levels (within populations, among populations within races, and among races). These distances were used by Cavalli-Sforza *et al.* (1994) to estimate evolutionary trees for human populations.

The forensic science community has interpreted the βs as though they were the θs needed for conditional genotype probabilities, and therefore has assigned the smallest values to θ for substructure within populations within racial groups instead of assigning the largest values. If the βs are to be used, care is needed with the choice of allele frequencies. Using data from subpopulations $S_1$, $S_2$ within the same population furnishes an estimate of $\beta_S$ that must be used with allele frequencies in the ancestral population $P_1$. Any conclusions therefore are conditional on those frequencies, and these are not knowable because they refer to a population no longer extant. Any analysis that is to make use of the coancestries θ instead of the βs is handicapped by lack of knowledge of the value of $\theta_A$. This ancestral coancestry cannot be estimated because there is no replicate of human populations, and it is usually assumed to be zero.

Here, then, is a limit to knowledge: It is not possible to estimate the absolute degree of relationship among the most closely related individuals unless an assumption is made that the most distantly related individuals are in fact unrelated. This is the essence of the "relative to" nature of Wright's *F* statistics. It is not yet clear if data from other species could be used in a similar way that out-groups are used to root phylogenetic trees.

## CONCLUSION

Evolution is a stochastic process so that past populations cannot be characterized with certainty. This makes it difficult to invoke population genetic theory to estimate the probability of a genetic profile in a population conditional on having seen the profile once already. The lack of specific historical information also makes it difficult to estimate quantities such as the recombination fraction from current population data.

ACKNOWLEDGMENTS

I thank members of the Program in Statistical Genetics at North Carolina State University. This work was supported in part by National Institutes of Health grant GM45344.

## REFERENCES

Balding, D. J., and Nichols, R. A., 1994, DNA profile match probability calculation: How to allow for population stratification, relatedness, database selection and single bands, *Forens. Sci. Int.* **64:**125–140.

Cavalli-Sforza, L. L., Menozzi, P., and Piazza, A., 1994, *The History and Geography of Human Genes*, Princeton University Press, Princeton, New Jersey.

Cockerham, C. C., 1973, Analyses of gene frequencies, *Genetics* **74:**679–700.

Evett, I. W., and Weir, B. S., 1998, *Interpreting DNA Evidence*, Sinauer, Sunderland, Massachusetts.

Hill, W. G., and Weir, B. S., 1994, Maximum-likelihood estimation of gene location by linkage disequilibrium, *Am. J. Hum. Genet.* **54:**705–714.

Kaplan, N. L., Hill, W. G., and Weir, B. S., 1995, Likelihood methods for locating disease genes in nonequilibrium populations, *Am. J. Hum. Genet.* **56:**18–32.

Ott, J., 1991, *Analysis of Human Genetic Linkage*, 2nd ed., Johns Hopkins Press, Baltimore.

Weir, B. S., 1996, *Genetic Data Analysis II*, Sinauer, Sunderland, Massachusetts.

Weir, B. S., and Hill, W. G., 1986, Nonuniform recombination within the human β-globin gene cluster, *Am. J. Hum. Genet.* **38:**776–778.

Wright, S., 1951, The genetical structure of populations, *Ann. Eugen.* **15:**323–354.

# IV

# Quantitative Genetics and the Prediction of Phenotype from Genotype

The translation from genotype to phenotype represents one of the most complex problems in biology. Classic quantitative genetics seeks to predict phenotypic change through single generations by imposing a highly simplified model of genetic determination on data obtained from specific mating designs. Least-squares analysis applied to these data maximizes the fit to an additive model of gene action. Interactions (intra or interlocus) are treated as a residual term. This model works well for short-term prediction, but it tells us nothing about the actual gene action that determines the character of interest. Both Clark and Lynch deal with this problem, but from rather different perspectives. Wu considers the problem of speciation in the context of gene interaction.

Clark (Chapter 11) takes human medical genetics, where the data are more extensive than in any other biological system, as a starting point in his exploration of the prediction of phenotype from genetic data. He makes an obvious but crucial distinction between medical genetics and agricultural applications of quantitative genetics by observing that in medical genetics the problem is to predict an individual phenotype, whereas in agriculture the problem is to predict changes in a population mean value. For the goal of predicting an individual phenotype, Clark lists several fundamental problems associated with the prediction of genetic risk in individuals, and he notes cases in which major mutations have failed to predict a significant proportion of the variance in risk in the population (e.g., *BRCA1* fails to account for the great majority of the risk for breast cancer).

Clark then goes on to ask a very important question: Can we understand the genetic basis of a complex phenotype from analyses of marginal effects abstracted from a system of much higher dimensionality? The motivation for this question is that in genetics we always observe a marginal system abstracted from an unobserved system of much higher genetic dimensionality. Does it matter if factors not observed in the experimental protocol interact with the factors actually measured? Clark shows by exploring the *NK* model of Stuart Kauffman that it can matter a great deal. Consideration of genotype × environment interaction also involves a complex system where inferences almost always are drawn from a subset of the total range of environmental possibilities. Clark notes that a system characterized by "complex and rich interactions among its components will in general appear to be simpler when viewed at a coarser level of resolution." It is unlikely that marginal analyses will accurately predict the behavior of a complex system with interacting components.

Lynch (Chapter 12) considers recent developments in quantitative genetics and asks whether these will provide a deeper understanding of the nature of adaptive variation and of the limits on rates of adaptive evolution. The development of quantitative trait locus analysis is discussed as a major advance in recent years, although this approach has two major limitations. The first is the inability to detect genes of small effect, and the second is a bias toward false-positives. Lynch asks whether existing quantitative variation is relevant to evolutionary adaptation or whether it is simply the residue of a mutation–selection balance. This question is fundamental to understanding the adaptive limits of population response, a crucial consideration in an era of increasing environmental change.

Wu (Chapter 13) considers the search for general rules in understanding speciation and the genetic nature of species differences. He discusses Haldane's rule and argues that students of speciation have been misled by a misplaced search for generality. He then address the problem of extrapolating from gross genetic measures of species differences to the underlying gene interaction systems that are responsible for reproductive divergence. Finally, he asks whether short-term studies of evolution are adequate guides to the processes of change that span millions of years.

The phenotype–genotype dichotomy is at the core of modern biological research. Evolutionary genetics approaches this complicated terrain from the methodological position of statistical inference. The existing statistical methodology is unlikely to be adequate to this task, thereby limiting our knowledge.

11

# Limits to Prediction of Phenotypes from Knowledge of Genotypes

ANDREW G. CLARK

## THE PROBLEM

The fact that natural selection acts on phenotypes but the transmission of traits to the next generation is indirectly accomplished through genes gives rise to a challenging set of problems in evolutionary biology. In order to understand adaptive evolution, it appears to be essential to first understand how genotypes give rise to observed phenotypes, or more precisely, how variation in phenotypes is mediated by underlying variation in genotypes. As the tools of molecular genetics give an increasingly detailed view of the underlying genetic variation, one would hope that this problem would be solved by the sheer volume of genetic data. Human molecular genetics has produced many significant successes recently, particularly in identifying genes that cause Mendelian disorders. In stark contrast, chronic diseases that exhibit familial clustering but do not segregate like a Mendelian gene have been remarkably difficult to analyze genetically. The focus of this chapter is on the question, "What are the barriers to our understanding of the genetic basis for familiar clustering of chronic diseases?" We will focus on medical genetics rather than the more general problem of genotype–phenotype associations in evolutionary biology, because

ANDREW G. CLARK • Institute of Molecular Evolutionary Genetics, Department of Biology, Pennsylvania State University, University Park, Pennsylvania 16802.

*Evolutionary Biology, Volume 32*, edited by Michael T. Clegg *et al.*
Kluwer Academic / Plenum Publishers, New York, 2000.

knowledge of phenotypic variation is so extensive for humans and the quantity of data on genetic variation is soon going to eclipse that of all other species, if it has not already.

Reading the mass media, one would surmise that there are no limits to genetic understanding of phenotypic variation. Genes for obesity, reading disability, aggression, happiness, and so forth are reported on a daily basis. This level of breathless excitement about the potential of genetics to understand the human condition should be a cause for great concern among serious geneticists. The stark reality is that genotype × environment interactions and other complications may make it almost impossible to be able to predict some phenotypes from genotypic information. It is time to ask what are the limits to the ability of genetics to predict phenotypes, and perhaps more importantly how the techniques that are applied to find genes for complex traits can give spurious answers. How serious is the risk that current technologies will continue to generate false-positives, and are there ways to identify which traits are likely to be recalcitrant to successful genetic analysis?

The first thing to emphasize is that our view of genetic reality is always obtained from a relatively small number of genes sampled from a relatively small number of individuals in a population. Our data represent a vast reduction in the number of dimensions that may actually have causal impact on phenotypes. All genetic research is really conducted on marginal effects drawn from a vastly high-dimensional state space. For example, a gene may have different effects in different environmental contexts, but the average of these effects might cancel out. Thus, the lack of a marginal effect does not mean that a gene has no effect on the phenotype, so it is important to attempt to identify situations in which a marginal picture is inconsistent or otherwise misleading.

Our understanding of the genetic basis for a trait may be represented by two very different types of models. In some cases we seek to understand how a gene functions by inference of the causal mechanism whereby an allele gives rise to a trait. Such a model arises from an understanding of the regulation of the gene and the biochemical consequences of the gene's product. Such a model is mechanistic, and this kind of understanding can suggest means by which the deleterious effects of a gene might be ameliorated. In other cases, a statistical model is constructed, which may have some ability to predict a phenotype given a genotype. Such a model need not entail any aspect of biological mechanism at all and need not even identify the function of the genes involved. Often for medical purposes, accurate prediction of disease risk is of great value, even if nothing is learned about causal mechanism.

There seem to be two extreme camps in genetics. One takes the view that simple statistical models have done a good job of describing variation in the past, so it is safe to ignore complexities like gene interactions and first see how well the simple models fit. The argument is that everyone knows that the world may be complex, but if it is too complex, we would not be able to make predictions anyway, so attempts to fit simpler models are the most fruitful. The second camp gets so worried about potential complexities that simple summaries are missed (see, for example, Chapter 5 in Levins and Lewontin, 1985). A question worth careful study is whether a complex reality can spuriously appear to fit simpler models. The statistical power to identify interaction effects is less than the power to identify main or marginal effects, primarily because of the difference in degrees of freedom (Sokal and Rohlf, 1995, p. 339), so it would seem that the tools we have immediately bias the picture toward a simpler-than-reality view (Sing *et al.*, 1996).

The reasons that genotype–phenotype predictions may be difficult to make are well known and should be taught in every first-year course in genetics. The first difficulty in identifying genes that affect a continuous distribution of risk is that there may be many genes of large effect, each of which by itself may cause a deleterious phenotype in only a small subset of the population. This is known as genetic heterogeneity, and congenital deafness is one of the standard textbook examples. To illustrate the problem, a Medline search on the keywords "congenital deafness" identifies 1,685 references with more than two dozen known genetic defects. The second barrier to identifying relevant genes is that the genetic basis may be too dispersive; that is, many genes all of very small effect combine to determine disease risk, and the loci that are causal differ from individual to individual. This is the standard infinitesimal model of quantitative genetics; while it makes for nice mathematics, if no single gene has a major effect on the phenotype, it is in practice very difficult to identify any of the causal genes. The third complicating factor is epistatic interactions, such that the effect on disease liability of an allelic substitution at one genetic locus is conditional on the state at another locus or loci. One might argue that so long as there is a significant marginal effect, such interactions do not matter, but they become important if different populations have different frequencies of genotypes at interacting loci. The fourth complication is genotype × environment interaction, whereby the effect of an allelic substitution on the phenotype varies depending on the environment. In the case of chronic disease phenotypes, the most relevant environmental variables are diet, smoking, stress, and perhaps exposure to pathogens. Although the best epidemiological methods attempt to account for interactions among causal variables, the sampling is such that the power to identify interactions is generally low. Let us examine each of these four complications in more detail, and see just how they impose limits to our ability to ascribe genetic causality to phenotypic variation.

## GENETIC HETEROGENEITY

Many chronic diseases are already known to be caused by defects in multiple genes. Cardiovascular disease risk is greatly increased by any change in lipid metabolism that results in elevated serum cholesterol, and lesions in any of several genes result in familial hypercholesterolemia (Lusis, 1988). Knowledge of metabolic pathways for regulation and transport of lipids has progressed, along with the identification of the molecular basis for these defects. But despite this progress, it is commonly assumed that only a small portion of the variance in serum cholesterol levels is caused by the genes whose defects result in familial disorders. The entire population may be considered to have a normal distribution of serum cholesterol levels, but the ascertainment for these genetic defects has been from clinical cases (i.e., extreme phenotypes). Current methods are very good at identifying the major genes for such extreme cases, and even if multiple genes are responsible for the diseases, they can eventually be identified. The subtlety comes in considering the distribution of risk of the rest of the population. It is quite possible that the genes that give rise to extreme phenotypes when major mutations occur are not the same genes that cause the bulk of the variation in risk in the general population. Most likely each causal gene exhibits a continuum of alleles with few and rare alleles of very large effect, and many alleles having small effects. The dichotomy between simple Mendelian factors and complex multifactorial disease is a bit arbitrary.

The approach of identifying genetic defects that are in common among groups of affected individuals has been very successful, but the above arguments make clear that more work is needed to determine how useful information about the identified gene is. The problem is that one would like to be able to predict risk, and lack of a defect in such a major gene only rules out a major defect and yields no more information about risk. Whether that gene explains much of the variation in a population requires a subsequent survey of the entire population. This is exactly what happened with the gene *BRCA1*. Defects in *BRCA1* clearly cause elevated risk of breast cancer, but the utility of a screening program was only recently tested by surveying samples from the general population and asking how many of the breast cancer cases had a defect in *BRCA1*. In one such sample, only 3 of the 211 white women with breast cancer had a *BRCA1* and none of the black women with breast cancer had a *BRCA1* defect (Newman *et al.*, 1998). In another independent study, 12 out of 193 women diagnosed as having breast cancer before age 35 years had *BRCA1* mutations (Malone *et al.*, 1998). Prospects for screening are dim because there are evidently many causes of breast cancer.

## MULTIPLICITY OF SMALL EFFECTS

More subtle problems arise when no major genes are identified as causal to risk. The etiology of an individual's disease then may be a cumulative effect of many genes, such that variants of no single gene can cause the disease. Human genetic methods have not been very powerful at identifying this class of gene, but quantitative trait loci (QTL) mapping methods have been widely applied in laboratory and agriculturally important plants and animals, and of course they often do identify segments of the genome that contribute to the variation in the phenotype. The ability to find all such genes is severely limited by statistical issues, including the relatively low power of the tests. Many of the designs make use of large numbers of genetic markers distributed throughout the genome. Often the number of markers is large relative to the number of potentially recombinant lines (or individuals) that are examined, and in this case the classical statistical problem of multicolinearity arises. In regression, if the sample size is small relative to the number of independent variables, then at least some of the independent variables are forced by this sampling to be correlated with one another (Neter *et al.*, 1983). If one looks at enough markers in the genome, some of them will be perfectly correlated, and simulations have shown that in this case spurious QTLs are identified (Beavis, 1994). The development of DNA chips for scoring large numbers of single nucleotide polymorphisms (SNPs) (Chee *et al.*, 1996) promises to make this problem quite serious unless care is taken in interpreting the results. Fortunately, methods are available to assess the significance of QTL effects in the face of this multicolinearity problem by performing permutation tests (Churchill and Doerge, 1994; Doerge and Churchill, 1996), but it is important to realize that this is a problem where more data do not necessarily produce more useful information.

The problem of multicolinearity of anonymous markers in mapping studies raises a particularly troublesome situation in human studies. One can easily imagine association studies of a few hundred individuals from each of two or more groups being compared for some complex phenotype, like reading disability, for example. If DNA chips are used to score, say, 10,000 SNPs, then spurious associations with the phenotype are nearly guaranteed. An $F_{ST}$ among Europeans and Africans has been estimated at approximately 0.15 for nuclear DNA sequences (Cavalli-Sforza *et al.*, 1994), so one would not be surprised to find fixed differences between these groups in this kind of sample. It is clear that biomedical researchers need to be educated on this issue, as the sampling aspects of the problem seem almost guaranteed to generate spurious fixed genetic differences between races associated with whatever socially relevant phenotype one chooses. Aside from the political bomb this situation represents, identification of true genetic cause for variation will be exceedingly difficult if many genes of small effect are involved. The challenge is to be able to identify this problem and arrive at a decision when to stop searching and consider the phenotype to be too complex to be worth further inquiry. Unfortunately, the problem of multicolinearity seems to guarantee significant effects for all but the statistically most prudent investigators.

## EPISTATIC INTERACTIONS

Epistasis plays an interesting role in evolutionary genetics. On the one hand, based on consideration of the complexities of the intermolecular interactions that occur to regulate gene expression and subsequent complexities of metabolism, it would appear that additivity of gene effects is highly unlikely at face value and that epistasis in this mechanistic sense must be ubiquitous (Gibson, 1996). On the other hand, classic quantitative genetics considers epistasis to be an interaction in a statistical model. These models have had and continue to have remarkable success in predicting selection response and familial resemblance through simple linear (additive) models. The success of classic quantitative genetics, which generally lumps epistatic interactions with the error variance, has served to encourage geneticists to believe that the simple linear models are adequate. But the model that provides a rough approximation of the behavior of the mean of a population may be far from adequate in predicting individual phenotypes. This distinction between mean effects and individual prediction is a fundamental difference between classic quantitative genetics and the goal of research in complex disease, although on the way to attempting individual prediction, genetic epidemiologists often build statistical models based on ensemble properties of the population. Despite the apparent need to be concerned about gene interactions, interactions are nearly universally ignored or assumed to be absent in these statistical models.

The perception that epistasis is relatively weak primarily comes from the tests applied by classic quantitative geneticists. Cockerham (1954) developed the methodology to partition variance into additive and dominance components and four additional components of epistatic interaction: additive × additive, additive × dominance, dominance × additive, and dominance × dominance. A pattern that emerged from such classical quantitative genetic approaches is that tests done with widely divergent lines generally show more epistasis than do closer crosses (Whitlock *et al.*, 1995). The classic quantitative genetic approach relies on standard linear models to partition variance, which means that the power to detect interactions is much lower than that for main effects, so that epistasis is essentially only seen if there also are marginal effects. There is no biological reason why epistatic interactions also must have single-locus marginal effects.

Many features of cell biology, such as the assembly of complex, multicomponent transcription factors, suggest that interactions of three and more components might be frequent. Tests of three-way epistasis are rarely attempted, in part because of the low statistical power of such a test. One example of such a test made use of balancer chromosome segregation data, and as might be expected from the diminishing power, significant higher-order interactions were less abundant than pairwise interactions but some cases of significance were found (Clark, 1987). It should be noted that the real world may in fact have many three- and four-way genetic interactions but that we simply do not have the tools to identify them.

Measured genotypes allow tests of epistasis that nearly match the statistical power of tests of additivity and dominance. Edwards *et al.* (1987) examined two sets of $F_2$ maize populations, scoring 20 markers and 82 traits, and although they found QTLs for every trait examined, only about 3% of tests of epistasis were significant. In general, measured genotype methods have revealed relatively little epistasis in any plant species (Tanksley, 1993). However, crosses between maize and teosinte have revealed strong epistasis in several characters (Doebley and Stec, 1991; Doebley *et al.*, 1995). In *Drosophila*, Long *et al.* (1995) report tests of pairwise interactions on bristle phenotypes among genomic regions marked by *roo* transposable element insertions. The design was an analysis of $F_2$ flies derived from crosses of high- and low-bristle number selected lines. Of the 60 tests that they could perform, 13 exhibited significant epistasis. They also found that the magnitude of epistatic effects was on par with the additive effects. In a QTL study of mouse body weight, Cheverud and Routman (1995) found that 15% of 120 two-locus tests yielded significant epistasis. It seems that the measured genotype approach provides a good tool for quantifying gene interactions for a variety of traits.

It also may be informative to consider the sorts of traits that tend to be more highly epistatic. At some level, it seems obvious that the determinants of an individual gene's expression must be highly nonadditive. The assembly of a transcription complex occurs only through delicate pairwise and multiway interactions among a large (and growing) set of transcription factors and related proteins. It follows that traits that are likely to be manifestations of expression of one or a few genes are likely to be epistatic because genetic variation in the factors that result in that gene's expression are likely to interact. Clark and Wang (1997) found extensive epistasis among *P*-element inserts that affect enzyme activities. Damerval *et al.* (1994) used two-dimensional gel electrophoresis to examine expression of 72 proteins in 85 $F_2$ maize plants and found that 14% of the proteins exhibited significant epistasis between genomic regions. Recall that morphological traits in maize have repeatedly exhibited little epistasis, yet traits close to the expression of individual genes appear to exhibit significant levels of epistasis. On the other hand, Stam and Laurie (1996) found that, apart from the *F/S* difference, no single change in *Adh* sequence had a large effect on ADH activity, but that several sites within the *Adh* region acted more or less additively to create alleles of larger effect. The observation of near additivity at this molecular scale was unexpected if one thinks in terms of the relevant molecular interactions; but even more importantly, the study shows that the effects identified by a QTL study may well entail many sites of small effect.

## SIMPSON'S PARADOX: COMPLEX SYSTEMS VIEWED FROM THE MARGINS

An important question is whether a highly epistatic system will appear highly epistatic when only a small subset of the relevant loci are examined and whether marginal effects must be present before an interaction can be detected. This is a biological statement of a question that arises in the study of multiway contingency tables: What can be said about the significance of multiway interactions based on observations of a table collapsed to fewer dimensions? The answer is that a marginal subsystem tells us very little about the full system. The easiest way to see this graphically is in Fig. 1, where the false correlation paradox is illustrated. In the figure, there are several ranges of X where within each range there is a strong negative correlation between X and Y. The four clusters of points may be measurements of X and Y taken in four different cities, for example. Despite the strong

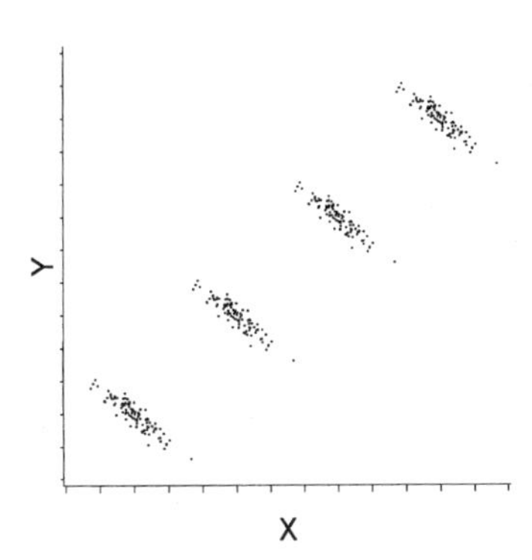


FIG. 1. Hypothetical example of a distribution of points that illustrates the paradox of false correlations. For each of four ranges of X there is a strong negative correlation, but overall, the data exhibit a positive correlation.

negative correlation within each city, the overall correlation of X and Y is strong and positive. In this case the paradox arises because the negative correlations are conditional on ranges of X that appear to differ significantly among cities.

Simpson's paradox is that tests of marginal effects in a multiway contingency table may be at odds with tests of the same variables conditioned on a third variable (Simpson, 1951). The paradox was illustrated by a hypothetical example giving the success rate of drug trials in two cities (Blyth, 1971). In the two cities the counts of cases in the various categories are as follows:

| | Chicago | | | Boston | |
|---|---|---|---|---|---|
| | *Control* | *Drug* | | *Control* | *Drug* |
| *Dead* | 950 | 9,000 | *Dead* | 5,000 | 5 |
| *Alive* | 50 (5%) | 1,000 (10%) | *Alive* | 5,000 (50%) | 95 (95%) |

These tables make it apparent that both cities show that the drug has a benefit, because the fraction of survivors is higher among the individuals who received the drug treatment than the controls. When the data are pooled together, collapsing over the two cities, something surprising happens:

| | *Control* | *Drug* |
|---|---|---|
| *Dead* | 5,950 | 9,005 |
| *Alive* | 5,050 (46%) | 1,095 (11%) |

Now it appears that the drug has a much lower survival rate overall. What happened? Your expectation that the collapsed table should have the same drug × survival relation is only true if the number of drug versus control cases is the same in the two cities. The complete $2 \times 2 \times 2$ table has a strong city × drug interaction because a much higher fraction of cases got the drug in Chicago. The hidden interaction results in a reversal of the apparent drug × survival interaction in the marginal table.

Is it ever possible to collapse a table and still have an accurate picture of the interactions between the remaining variables? Bishop *et al.* (1975) prove a theorem about collapsing a three-dimensional contingency table, which states that the interaction between the two remaining variables is accurately captured in the collapsed table, provided at least one of the two variables has zero interaction with the variable that is collapsed. An analogue to this theorem can be seen in the formula for partial correlations. Letting $\rho_{12.3}$ be the partial correlation between variables 1 and 2, conditioned on variable 3, and $\rho_{12}$, $\rho_{13}$, and $\rho_{23}$ be the simple correlation coefficients, we have

$$\rho_{12.3} = \frac{\rho_{12} - \rho_{13}\rho_{23}}{\sqrt{(1-\rho_{13}^2)(1-\rho_{23}^2)}}$$

This formula shows that if either $\rho_{13} = 0$ or $\rho_{23} = 0$, then $\rho_{12.3}$ will be proportional to $\rho_{12}$, so the hypothesis test that variables 1 and 2 are correlated will be equivalent. The only problem with this conclusion is that it means that we have to know about interactions of variables 1 and 2 with variable 3, and if variable 3 is unknown, we can never be sure whether our test of interaction of variables 1 and 2 will be affected by another unmeasured variable.

Simpson's paradox makes clear that observation of strong interactions in a complete table do not guarantee that collapsed tables will manifest the same interaction (Agresti, 1990). This is relevant to genetic problems because all experiments and observations that we make in genetics are performed on the margins. We cannot look at all genes, so we choose some subset of genes to examine. When we claim that there are no detectable interactions in the marginal subsystem that we measure, we have no grounds at all to make conclusions about interactions with genes that are not observed. Furthermore, the interactions between the genes that we do observe may be a spurious consequence of pooling across unobserved interaction genes. This should be a bit unsettling.

## THE *NK* MODEL VIEWED AT THE MARGINS

In order to study the behavior of highly epistatic models, Stuart Kauffman devised a model in which $N$ loci could each interact with $K$

other loci to produce a net fitness (Kauffman, 1993). The model is meant to be conceptual, allowing a rich set of potential interactions among genetic loci. For example, because the model can allow high-order interactions, it accommodates the possibility that some genotypes may have loci 1 and 2 interact to determine a trait, whereas other genotypes may have additivity over loci 1 and 2 but instead show interaction between loci 3 and 4. The model is haploid and disallows recombination, but there is no reason why extensions to diploidy and recombination could not be added. Considering the haploid, biallelic case, the $N$ loci allow $2^N$ haplotypes. If each locus interacts with $K$ other loci, $2^{K+1}$ values of $w_i$ are chosen from a uniform distribution between 0 and 1. The net fitness of a genotype is

$$w = \frac{1}{N}\sum_{i=1}^{N} w_i$$

Kauffman made several observations of the $NK$ model that bear repeating. Considering the extreme cases first, if $K = 0$, the model has no interlocus interactions, so fitness is simply the sum of effect of individual loci. In this case the fitness surface has a unique maximum and the population inexorably climbs to this fitness peak. Kauffman gives this situation the colorful name Fujiyama. If $K = N - 1$, every locus interacts with every other, so the knowledge of the fitness of one genotype tells us nothing about the fitness of a genotype one mutational step away. Such a system has many local fitness maxima, a situation Kauffman calls a rugged fitness landscape. In such a case, the fate of the population depends critically on initial conditions. In addition to these fairly obvious properties, some interesting things emerge from simulations of the $NK$ model.

Let us construct samples from the $NK$ model and ask a very simple question. Knowing the full model, we can determine whether an allelic substitution at one locus will initially increase or decrease in frequency; it depends only on whether the average fitness of that substitution is greater or lesser than the mean fitness in the population. After determining whether an allele will increase or decrease, we then consider some subset of loci $x$, where $x < N$, and ask whether in this marginal system we would make the same prediction.

In the case of $K = 0$, there is no interaction, so the fate of an allele depends only on its own effect on fitness. In such a case, the marginal system must behave like the full system, and the simulations verify this (Fig. 2). For $K = N - 1$, we have a maximally interactive system, such that each gene interacts with every other gene. In this case, if all genes are examined, we have perfect knowledge, so prediction of phenotype is 100% accurate. But as soon as a single locus is not observed, we lose all information about phenotype, so the fate of an allele is like a random coin toss and we are correct

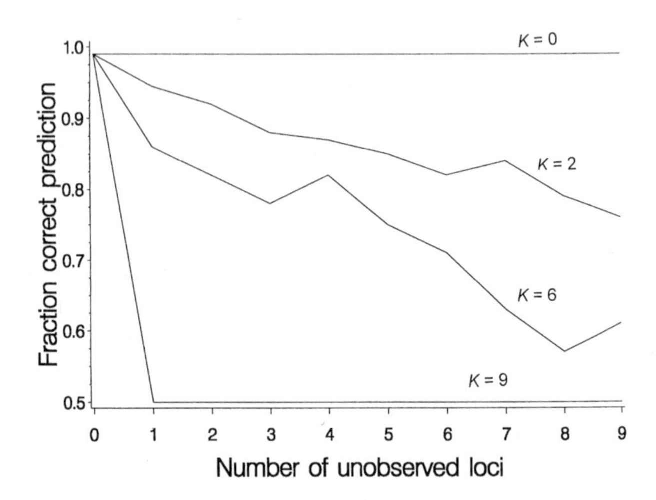


FIG. 2. Accuracy of prediction of whether an allele will increase in the Kauffman *NK* model based on an observed subset of loci. The full model has 10 loci, with each locus interacting with $K = 1, 2, \ldots 9$ other loci, and the observed subsystem may have fewer loci. For each trial, we ask whether an introduced allele would increase in the full system, and whether the subsystem would accurately predict initial increase of the allele.

only half the time. For $K = 2$ or $K = 6$, the situation is intermediate. $K = 2$ is more tolerant of missing loci than is $K = 6$, because the accuracy in predicting phenotype is greater for a given number of unobserved loci. This means that more interactive systems yield less reliable information in the margins. An important point to emphasize here is that we do not know where reality lies. The fact that genotype–phenotype associations appear to reverse direction or involve different genes in different populations suggests that at the very best we are in an intermediate state of ignorance for most phenotypes.

## *N*-LOCUS DIPLOID EPISTASIS VIEWED FROM THE MARGINS

The haploid *NK* model allows extreme levels of interaction among loci but still suffers from being haploid. The epistatic model of Zhivotovsky and Gavrilets (1992) is explicitly diploid but considers only pairwise interactions. Because it allows one to construct samples from a population with an arbitrary number of interacting loci, it too warrants examination of marginal properties. Let $g_i$ and $g'_i$ be the allele indices (0 or 1) for the maternal and paternal allele at locus $i$. Fitness is determined as follows:

$$w = \mu + \sum_i [a(g_i + g'_i) + b g_i g'_i)] + c \sum_{i?j} (g_i + g'_i)(g_j + g'_j)$$

where $a$ is the additive effect, $b$ is the dominance effect, and $c$ is a single parameter that determines the magnitude of additive × additive epistasis. The first thing to note about this model is that even when $a = 0$ and $b = 0$, so long as $c \neq 0$, there is still additive variance in the classic quantitative genetic sense. Partitioning of the total genetic variance reveals an additive component, even when the model is generated with pure interaction. This case is illustrated by generating random 100-locus genotypes for 100 families with two parents and four offspring. Both algebraic decomposition of the variance and simple simulations of parent–offspring regression show that even with $a = b = 0$, and $c = 1$, there is considerable heritability of this epistatic trait (Fig. 3). A strong conclusion is that just because a trait exhibits a nicely behaved normal distribution and a strong parent–offspring regression, this does not mean that there is zero functional epistasis. One of the initial claims that epistasis is perceived to be weak is based on just this notion: If statistical models fit adequately without invoking epistasis, then the underlying biology must in fact lack interaction among components. Such a conclusion is clearly wrong. The Zhivotovsky and Gavrilets model

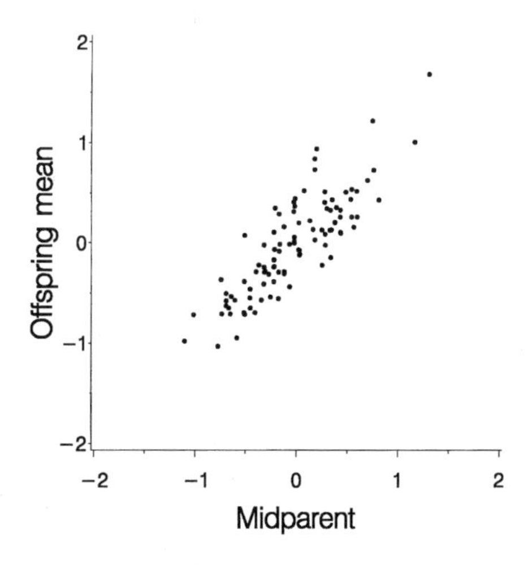


FIG. 3. Parent–offspring scattergram simulated with the Zhivotovsky and Gavrilets (1992) model with $a = 0$, $b = 0$, and $c = 1$. Despite the high level of pairwise epistasis, the trait appears to have high additive variance and high heritability.

shows that additive statistical models can fit data very well even in the face of strong underlying epistasis.

Now consider what happens to the Zhivotovsky and Gavrilets model when we examine only a subset of loci. In the limiting case of no recombination, an allele will increase in frequency if the marginal fitness of the allele, defined as $w_i = \Sigma\Sigma\Sigma \ldots p_i p_j p_k \ldots w_{ijk} \ldots$ is greater than the mean fitness of the population. Samples of 100-locus genotypes in a Hardy–Weinberg population were generated, and fitnesses of each genotype were calculated under the constraint $b = 0$, $a + c = 1$. This model allows for a trading off between additive and epistatic effects. A locus and allele were chosen at random, and it was determined from the full model whether the allele would increase in frequency. Then different numbers of loci were hidden from view and the marginal and mean fitnesses were recalculated in order to determine whether the nested model gave the same prediction of allele frequency change. Results again show that the more highly interactive a model is, the more sensitive is the assessment of allele frequency change to subsampling of loci (Fig. 4). In the purely additive case, all that is needed is the target locus, and its dynamics are correctly predicted. As interaction becomes stronger, the accuracy in predicting allele frequency change requires ascertainment of more loci.

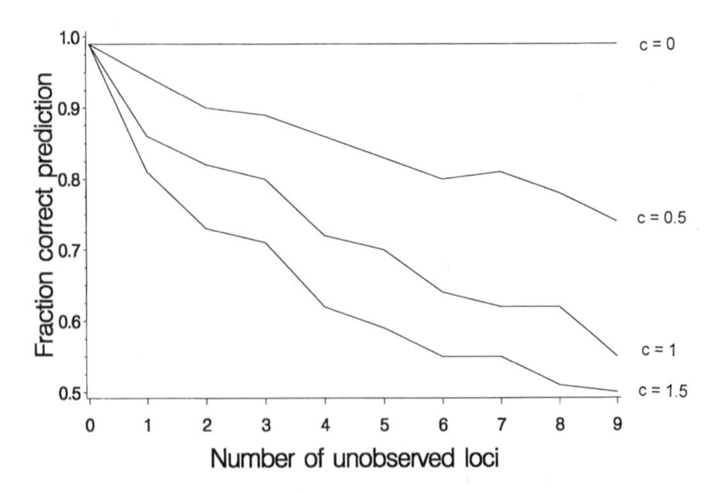


FIG. 4. Plot of accuracy of prediction of whether an allele will increase for the diploid Zhivotovsky and Gavrilets (1992) model. Layout of the plot is like that in Fig. 2, but this model only allows pairwise interactions.

In the above analysis we asked how well the fate of an allele can be predicted from partial knowledge of the genotype. Now consider tests of two-locus epistasis embedded in a model having higher-order interactions. How often do we see apparent additivity in this lower-dimensional view when in fact there is high-order interaction? This can be examined with both the *NK* model and the model of Zhivotovsky and Gavrilets. Only preliminary results have been obtained, and they suggest that by dropping loci from the analysis, we are in effect collapsing the data into a marginal view. Such a collapsed system behaves like the average of the systems conditional on each allele at the loci that are collapsed unless there are interactions among the collapsed loci. We illustrated this above in the discussion of Simpson's paradox. Unfortunately, this means that there is no consistent tendency for a collapsed system either to reduce nor increase the apparent strength of epistasis compared to the full-dimensional system.

In experimental population genetics there is a long history of awareness of the effects of genetic background, particularly in tests of fitness components. The two approaches that have been taken to control for background genetic effects have been either to randomize the background by intercrossing lines or to isolate one or a few genetic backgrounds with balancer chromosomes. By isolating several genetic backgrounds, one at least has the opportunity to test for a background effect, but the sampling of backgrounds is generally very small. By randomizing over backgrounds, one loses the ability to test the significance of a background effect, but it is argued that the marginal result that one does quantify is likely to be that seen in nature. If experiments reveal apparent marginal additivity, the short-term behavior will be the same whether or not there is epistasis.

Concern about the validity of reducing dimensions of a highly complex system to a few tractable dimensions appears in many branches of science. Community ecology is another field in which one knows at the outset that it is impossible to quantify all relevant variables, so one can ask, "Under what conditions do we expect a subset of variables to reflect accurately what is going on in the full system?" Schaffer (1981) developed what he called abstracted growth equations to describe changes in density of a subset of a community, and he found some conditions under which stability properties of full systems and subsystems are consistent. This theoretical work also identifies experimental manipulations that can be done to alert the empiricist to important interactions with unobserved species. But in the end, Simpson's paradox reigns, and subsystems that may appear stable when they are actually not stable in the full model (and vice versa).

## GENOTYPE × ENVIRONMENT INTERACTION

The phenotype manifested by each genotype must depend on the rearing environment, and the fact that different genotypes respond differently to different environments is generally referred to as genotype × environment interaction. Among the most ambitious tests of genotype × environment interaction were QTL mapping studies carried out in multiple environments. Such studies readily identify particular regions of the genome that have effects on the phenotype in environmentally context-dependent ways (Stuber *et al.*, 1992; Xiao *et al.*, 1995). The near ubiquity of genotype × environment interaction means that any study done in a single environment produces conclusions that, strictly speaking, are limited to that environment. The relevance of genotype × environment interaction to the question of limits to knowledge of genotype–phenotype associations is closely related to that for epistasis. Unless we have full control over environmental contexts, our ability to make predictions of phenotypes will be compromised.

One example from the human genetics literature serves to illustrate the complexity of the problem. Apolipoprotein E is a remarkable locus in that alternative genotypes have a clear difference in serum cholesterol levels. This has attracted so much attention that extensive data now are available on the phenotypes of *ApoE* genotypes, enough data to stratify the sample and ask about subgroups. It turns out that there are enormous sex × genotype interactions, such that females exhibited no significant difference among *ApoE* genotypes in adjusted measures of total cholesterol, low-density-lipoprotein cholesterol, and low density lipoprotein, whereas males exhibited significant *ApoE* genotypic associations with all these traits (Kaprio *et al.*, 1991; Reilly *et al.*, 1991; Haviland *et al.*, 1995). Furthermore, the pattern of correlations among lipid components differed significantly among sexes and *ApoE* genotypes. Finally, serum lipid phenotypes are known to change with age, and Zerba *et al.* (1996) showed that the patterns of sex × genotype interaction changed significantly with age (i.e., sex × genotype × age interactions). General statements about the effect of *ApoE* on serum cholesterol that ignore the effects of age and sex now seem hopelessly naïve.

Referring again to Fig. 1, if each cluster of points represented a different environment, one could have a situation in which there is a negative correlation within each environment, but collapsing over environments would give the impression of an overall positive correlation between variables. If the X axis were an allele frequency and the Y axis a phenotype, the false correlation paradox would result in a reversal of the genotype–phenotype relationship. Similarly, the apparent interaction between two genes can be profoundly affected by pooling over environments. One could have zero epistasis in each of two environments, but strong apparent epistasis when pooled:

| Environment 1 | | | Environment 2 | | | Pooled | | |
|---|---|---|---|---|---|---|---|---|
| | *B* | *b* | | *B* | *b* | | *B* | *b* |
| *A* | 4 | 4 | *A* | 1 | 5 | *A* | 5 | 9 |
| *a* | 4 | 1 | *a* | 5 | 5 | *a* | 9 | 6 |

It is also possible to have strong epistasis in each of two environments, but the epistasis would not be apparent in the pooled data:

| Environment 1 | | | Environment 2 | | | Pooled | | |
|---|---|---|---|---|---|---|---|---|
| | *B* | *b* | | *B* | *b* | | *B* | *b* |
| *A* | 1 | 4 | *A* | 4 | 1 | *A* | 5 | 5 |
| *a* | 4 | 1 | *a* | 1 | 4 | *a* | 5 | 5 |

In sum, knowledge about a genotype–phenotype relationship in one environment does not allow inference about the genotype–phenotype environment in another environment.

## RECOMMENDATION

It is important to understand that an underlying mechanism that is complex and rich in interactions among its components in general will appear to be simpler when viewed at a coarser level of resolution. In the case of genotype × environment interaction, for example, studying the problem in one environment may make it appear that a gene has a simple effect on the phenotype. But another study in another environment also may give the appearance of a simple but different effect. Knowing that genotype × environment interactions can occur, what is the best experimental strategy? Are we to do all studies across a range of environments? The same situation arises with epistasis; examination of a trait in one set of genetic backgrounds may give different results from another. Even if the same genetic backgrounds are used, observation of different sets of markers may give different results. We cannot examine all genes in all backgrounds, so it is important to develop rules for identifying these problems.

ACKNOWLEDGMENTS

I thank Charlie Sing and Gary Churchill for discussions that helped develop some of the ideas in this chapter and for thoughtful comments on the manuscript. Support of National Institutes of Health grant HL58240 is acknowledged.

## REFERENCES

Agresti, A., 1990, *Categorical Data Analysis*, John Wiley & Sons, New York.

Beavis, W. D., 1994, The power and deceit of QTL experiments: Lessons from comparative QTL studies, in: *Proceedings of the 49th Annual Corn and Sorghum Industry Research Conference*, pp. 250–266, American Seed Trade Association, Chicago.

Bishop, Y. M. M., Fienberg, S. E., and Holland, P. W., 1975, *Discrete Multivariate Analysis: Theory and Practice*, MIT Press, Cambridge, Massachusetts.

Blyth, C. R., 1971, On Simpson's paradox and the sure-thing principle, *J. Am. Stat. Assoc.* **67:**364–367.

Cavalli-Sforza, L. L., Menozzi, P., and Piazza, A., 1994, *The History and Geography of Human Genes*, Princeton University Press, Princeton, New Jersey.

Chee, M., Yang, R., Hubbell, E., Berno, A., Huang, X. C., Stern, D., Winkler, J., Lockhart, D. J., Morris, M. S., and Fodor, S. P., 1996, Accessing genetic information with high-density DNA arrays, *Science* **274:**610–614.

Cheverud J. M., and Routman, E. J., 1995, Epistasis and its contribution to genetic variance components, *Genetics* **139:**1455–1461.

Churchill, G. A., and Doerge, R. W., 1994, Empirical threshold values for quantitative trait mapping, *Genetics* **138:**963–971.

Clark, A. G., 1987, A test of multilocus interaction in *Drosophila melanogaster*, *Am. Nat.* **130:**283–299.

Clark, A. G., and Wang, L., 1997, Epistasis in measured genotypes: *Drosophila P*-element insertions, *Genetics* **147:**157–163.

Cockerham, C. C., 1954, An extension of the concept of partitioning hereditary variance for analysis of covariances among relatives when epistasis is present, *Genetics* **39:**859–882.

Damerval, C., Maurice, A., Josse, J. M., and de Vienne, D., 1994, Quantitative trait loci underlying gene product variation: A novel perspective for analyzing regulation of genome expression, *Genetics* **137:**289–301.

Doebley, J., and Stec, A., 1991, Genetic analysis of the morphological differences between maize and teosinte, *Genetics* **129:**285–295.

Doebley, J., Stec, A., and Gustus, C., 1995, *Teosinte branched 1* and the origin of maize: Evidence for epistasis and the evolution of dominance, *Genetics* **141:**333–346.

Doerge, R. W., and Churchill, G. A., 1996, Permutation tests for multiple loci affecting a quantitative character, *Genetics* **142:**285–294.

Edwards, M. D., Stuber, C. W., and Wendel, J. F., 1987, Molecular-marker facilitated investigations of quantitative trait loci in maize. I. Numbers, genomic distribution and types of gene action, *Genetics* **116:**113–125.

Gibson, G., 1996, Epistasis and pleiotropy as natural properties of transcriptional regulation, *Theor. Pop. Biol.* **49:**58–89.

Haviland M. B., Lussier-Cacan, S., Davignon, J., and Sing, C. F., 1995, Impact of apolipoprotein E genotype variation on means, variances, and correlations of plasma lipid, lipoprotein, and apolipoprotein traits in octogenarians, *Am. J. Med. Genet.* **58:**315–331.

Kaprio, J., Farrell, R. E., Kottke, B. A., Kamboh, M. I., and Sing, C. F., 1991, Effects of polymorphisms in apolipoproteins E, A-IV, and H on quantitative traits related to risk for cardiovascular disease, *Arterioscler. Thromb.* **11:**1330–1348.

Kauffman, S. A., 1993, *Origins of Order: Self-organization and Selection in Evolution*, Oxford University Press, Oxford, England.

Levins, R., and Lewontin, R., 1985, *The Dialectical Biologist*, Harvard University Press, Cambridge, Massachusetts.

Long, A. D., Mullaney, S. L., Reid, L. A., Fry, J. D., Langley, C. H., and Mackay, T. F. C., 1995, High resolution mapping of genetic factors affecting abdominal bristle number in *D. melanogaster*, *Genetics* **139:**1273–1291.

Lusis, A. J., 1988, Genetic factors affecting blood lipoproteins: The candidate gene approach, *J. Lipid Res.* **29:**397–429.

Malone, K. E., Daling, J. R., Thompson, J. D., O'Brien, C. A., Francisco, L. V., and Ostrander, E. A., 1998, BRCA1 mutations and breast cancer in the general population: Analyses in women before age 35 years and in women before age 45 years with first-degree family history, *J. Am. Med. Assoc.* **279:**922–929.

Neter, J., Wasserman, W., and Kutner, M. H., 1983, *Applied Linear Regression Models*, Richard D. Irwin, Inc., Homewood, Illinois.

Newman, B., Mu, H., Butler, L. M., Millikan, R. C., Moorman, P. G., and King, M. C., 1998, Frequency of breast cancer attributable to BRCA1 in a population-based series of American women, *J. Am. Med. Assoc.* **279:**915–921.

Reilly, S. L., Farrell, R. E., Kottke, B. A., Kamboh, M. I., and Sing, C. F., 1991, The gender-specific apolipoprotein E genotype influence on the distribution of lipids and apolipoproteins in the population of Rochester, MN. I. Pleiotropic effects on means and variances, *Am. J. Hum. Genet.* **49:**1155–1166.

Schaffer, W. M., 1981, Ecological abstractions: The consequences of reduced dimensionality in ecological models, *Ecol. Monog.* **51:**383–401.

Simpson, E. H., 1951, The interpretation of interaction in contingency tables, *J. Roy. Stat. Soc. Ser. B* **13:**238–241.

Sing, C. F., Haviland, M. B., and Reilly, S. L., 1996, Genetic architecture of common multifactorial diseases, in: *Variation in the Human Genome* (K. M. Weiss, ed.), pp. 211–229, John Wiley & Sons, Chichester, England.

Sokal, R. R., and Rohlf, F. J., 1995, *Biometry*, 3rd ed., W. H. Freeman and Co., New York.

Stam, L. F., and Laurie, C. C., 1996, Molecular dissection of a major gene effect on a quantitative trait: The level of alcohol dehydrogenase expression in *Drosophila melanogaster*, *Genetics* **144:**1559–1564.

Stuber, C. W., Lincoln, S. E., Wolff, D. W., Helentjaris, T., and Lander, E. S., 1992, Identification of genetic factors contributing to heterosis in a hybrid from two elite maize inbred lines using molecular markers, *Genetics* **132:**823–839.

Tanksley, D. S., 1993, Mapping polygenes, *Annu. Rev. Genet.* **27:**205–233.

Whitlock, M. C., Phillips, P. C., Moore, F. B. G., and Tonsor, S. J., 1995, Multiple fitness peaks and epistasis, *Annu. Rev. Ecol. Syst.* **26:**601–629.

Xiao, J., Li, J., Yuan, L., and Tanksley, S. D., 1995, Dominance is the major genetic basis of heterosis in rice as revealed by QTL analysis using molecular markers, *Genetics* **140:**745–754.

Zerba, K. E., Ferrell, R. E., and Sing, C. F., 1996, Genotype–environment interaction: Apolipoprotein E (*ApoE*) gene effects and age as an index of time and spatial context in the human, *Genetics* **143:**463–478.

Zhivotovsky, L. A., and Gavrilets, S., 1992, Quantitative variability and multilocus polymorphism under epistatic selection, *Theor. Pop. Biol.* **42:**254–283.

# 12

# The Limits to Knowledge in Quantitative Genetics

MICHAEL LYNCH

## INTRODUCTION

The limitations to our understanding of evolutionary genetic phenomena lie primarily in the empirical domain, as mathematicians have repeatedly shown that they are up to the challenges of modeling essentially any population genetic phenomenon that can be envisioned. Until recently, one of the major constraints in our attempts to understand the mechanisms by which populations evolve has been the inaccessibility of the gene. Advances in molecular technology now enable us to routinely survey populations for variation at the molecular level, although almost all such surveys involving functional genes are focused on coding regions as opposed to the frequently much more complex (and potentially more significant) regulatory regions. The latter shortcoming will certainly be surmounted in the near future, and it is fair to say that no longer are there really any fundamental limitations (other than financial ones) to our ability to monitor the dynamics of individual alleles in natural or artificial populations. Nevertheless, despite the major advances in mathematical theory and in molecular technology, one might argue that we are not much closer to a mechanistic understanding of the evolution of complex phenotypes than we were in 1920. One of the major challenges confronting evolutionary geneticists is still the development of a general and biologically based synthesis that will facilitate such understanding.

MICHAEL LYNCH • Department of Biology, University of Oregon, Eugene, Oregon 97403.

*Evolutionary Biology, Volume 32*, edited by Michael T. Clegg *et al.*
Kluwer Academic / Plenum Publishers, New York, 2000.

I have chosen to restrict my discussion on the limits to knowledge in evolutionary biology to issues in quantitative genetics, the subdiscipline that deals with the inheritance and evolution of multilocus traits. It was interest in phenotypic diversity that drove the development of the so-called modern synthesis in the first half of this century, and the primary research of many evolutionary geneticists is still focused on the mechanisms responsible for morphological, physiological, and/or behavioral diversification within and among species. The vast majority of characters falling into these categories are quantitative, encoded by a large number (dozens to perhaps hundreds) of loci. What follows is a brief exposition on some of the major unresolved issues in quantitative genetics, their significance, and an attempt to address the degree to which they will ever be resolvable. History has shown that scientists regularly have surpassed barriers that were previously perceived as insurmountable, so one might argue that any attempt to identify the limits to scientific knowledge is a futile exercise. However, uncritical acceptance of dominant paradigms can impede scientific inquiry and progress. A close look at our perceived limits of knowledge and their relative significance may provide a useful perspective on the most useful directions of future research.

## PHENOTYPES VERSUS GENETIC VALUES

A serious limitation of the classic theory of quantitative genetics is its phenomenological nature. Because of the multilocus nature and environmental dependence of the expression of most quantitative traits, it is usually impossible to use phenotypic information to make precise statements about the underlying genotypes of individuals. As a consequence, most studies in quantitative genetics concentrate on variation at the population level and the partitioning of such variation into causal components (associated, for example, with additive effects, maternal effects, developmental noise, etc.) Sophisticated statistical methods [e.g., best linear unbiased prediction (BLUP)] do provide a basis for estimating the breeding values of individuals when phenotypic measures are available from multiple relatives, and these methods are employed extensively in animal breeding to identify elite individuals for selective breeding. In addition, for crop plants that can be propagated clonally or as inbred lines, general combining abilities can be estimated for individual lines and specific combining abilities for pairs of lines. However, all of these concepts are statistical, not biological, in nature. For example, a breeding value is simply an estimate of the expected phenotype of an offspring when the focal parent is randomly mated. As an estimate, a breeding value is never known with absolute certainty, but more significantly, such a composite measure tells us nothing about the actual allelic states of the loci involved in the expression of the phenotype or about modes of gene action.

Similar criticisms can be levied against the concept of variance component partitioning. For example, the additive genetic variance, which is the goal in attempts to estimate heritabilities of quantitative traits, is formally related to the variance of breeding values. Here, the word "additive" can be quite misinformative, as it refers to the behavior of phenotype distributions in a statistical sense and need not convey any information on the physiological mode of gene action (Cheverud and Routman, 1995). Situations exist in which there can be substantial nonadditive interactions among segregating genes and yet very little nonadditive genetic variance; the outcome depends on the average effects of alleles in all genetic backgrounds. As a consequence of the hierarchical way in which genetic components of variance are defined in a statistical sense, higher-order components involving nonadditive interactions are virtually always smaller in magnitude than the first-order (additive) component, and this appears to have led to widespread misconception that dominance and epistasis are of minor importance in the expression of quantitative traits.

Despite these limitations, quantitative genetics theory has been a fundamental source of many of the methods of applied statistics and has played an instrumental role in applications of selective breeding and in the development of management strategies in conservation biology. One of the great contributions of quantitative genetics is its ability to reveal the extent to which standing variation for complex traits is due to genes versus environment, but one of its greatest weaknesses is its failure to yield much insight into the biological nature of quantitative variation. There is now great hope that quantitative trait loci (QTL) analysis (the localization and characterization of QTL by use of molecular markers) will remove this shortcoming, an issue that I will address below.

## THE NATURE OF QUANTITATIVE VARIATION

Quantitative trait loci analysis involves the search for associations between molecular markers and the expression of quantitative traits. The degree of such associations depends on the magnitude of linkage disequilibrium between the marker locus and the QTL, on the rate of recombination between the two, and on the magnitude of the QTL effect. The very nature of the enterprise is biased in that loci with very small effects are generally undetectable and in that the estimated effects associated with detectable loci tend to be inflated (i.e., there is a statistical bias toward false-positives). Moreover, there generally is a very large gap between the crude localization of a QTL with a marker and its subsequent identification and characterization.

An alternative entree into the genetic basis of quantitative variation is the candidate locus approach. Here the idea is to exploit the rich array of studies in molecular–developmental biology that have revealed the functions of various protein-coding loci through the phenotypic effects of null mutations. Loci that are functionally connected to the trait of interest can then be surveyed for segregating variation involving more minor mutations in natural populations. This approach was successfully used to show that some of the variation for bristle numbers in natural populations of *Drosophila melanogaster* is associated with the *achaete-scute* region (Mackay and Langley, 1990) and the *scabrous* locus (Lai *et al.*, 1994). In addition, life history variation in *D. melanogaster* has been found to be associated with variation in copy number of the 28S ribosomal RNA gene (Templeton *et al.*, 1989; Hollocher *et al.*, 1992). These kinds of results raise the possibility that the loci whose major effects have been identified by knockout mutations in laboratory species may be the same loci that are responsible for smaller-scale variation in natural populations. If this is found to be the case, the field of inquiry in quantitative genetics will be dramatically expanded and facilitated.

We still are a very long way from a complete elucidation of the molecular basis of any quantitative trait, although the efforts now being expended in agricultural companies and in medical genetics suggest that major advances soon will be made. For example, substantial progress has been made recently in identifying the loci that are responsible for the major morphological differences between domesticated maize and its wild progenitor teosinte (Doebley and Stec, 1993). Thus, in principle, with enough effort, QTL analysis can reveal the full array of loci, alleles, and the interactions that determine the range of variation for specific quantitative traits in specific model systems. What then?

Although this line of endeavor has by no means reached the point at which we can clearly delineate the limits to what we can accomplish, the issues certainly can be addressed. It should be clear that QTL analysis at best reveals the loci that are currently expressing variation in a population. While such information may provide insight into future potential directions of evolution in a focal population, it does not necessarily tell us anything about the pathway by which the current population mean phenotype was acquired. The loci that were most involved in past evolution could very well be the ones lacking in observable standing variation. Moreover, it is by no means clear that the loci observed as significant in any one species have the same significance in other species.

One of the more remarkable generalities that has emerged from molecular studies of development is the degree to which homologous genes have been recruited for different functions in different species and the extent to which similar morphological structures in different species can be products of very different genetic pathways. Most of this work has involved comparisons of model species from different phyla (Gerhart and Kirschner, 1997), but recent work with nematodes has revealed how remarkably different developmental mechanisms can yield nearly identical morphological structures in species from different families (Eizinger and Sommer, 1997), and work with members of different orders of echinoderms has revealed similar complexities (Lowe and Wray, 1997). If these types of results turn out to be general, then the specific genes that are involved in morphological development in any one species may be fairly restricted in their phylogenetic relevance.

At issue here is the fundamental question of the extent to which information on microevolutionary processes yields any insight into macroevolutionary patterns. Arguments have been made that microevolutionary theory can explain adequately virtually all macroevolutionary phenomena (Charlesworth *et al.*, 1982), but it is not immediately apparent how the recent revelations in developmental biology can be accommodated by the existing theory. Certainly, it is fair to say that there is a large gap between our understanding of how bristle numbers evolve in natural populations of flies and of how the extraordinary degree of morphological differentiation arose among the animal phyla.

## THE MEANING OF QUANTITATIVE VARIATION

I now turn to the more fundamental issue of genetic variation itself and its relevance to studies in evolutionary genetics. One of the central parameters of quantitative genetics is the so-called narrow-sense heritability, which is the fraction of phenotypic variation that is additive genetic in basis. Practical interest in this parameter arises from the fact that the response to directional selection is expected to equal the product of the selection differential (the difference between mean phenotypes before and after selection, but prior to reproduction) and the heritability. This "breeders' equation" is the closest thing that we have to a predictive theory of evolution: If we know the heritability of a character, then we should be able to predict the response to a specific amount of selection. [Things get more complicated when multiple characters are simultaneously exposed to selection, but these complications can be accomodated by a multivariate form of the theory (Hazel, 1943; Lande, 1988).] One difficulty with the heritability concept is that methods for estimating it often only give crude approximations, as these are generally contaminated by nonadditive or nongenetic sources of variance. A more fundamental problem concerns the nature of standing genetic variation for quantitative traits.

A recent survey for a diversity of traits in a variety of plant and animal species suggests that the average persistence time of a new mutation for a quantitative trait is on the order of a hundred generations (Houle *et al.*, 1996). Since the reciprocal of the persistence time is approximately equal to the average selection coefficient against a mutant allele (Crow, 1992), this suggests that the average mutant allele for a quantitative trait has a deleterious fitness effect of about 1%. This level of selection intensity is qualitatively in accord with results that have been obtained in laboratory mutation accumulation experiments, raising the hypothesis that most of the standing genetic variation for quantitative traits is a simple consequence of a balance between the input of mildly deleterious alleles by mutation and their elimination by weak selection. If this is the case, then a substantial fraction of the standing genetic variation that we quantify as heritability may be of little relevance to long-term adaptive evolution. If the subset of alleles that are of relevance to adaptive evolution are essentially "rubies in the rubbish," then our usual surrogate measure of evolvability (the total standing amount of additive genetic variance) is misguided. Discriminating between genes that are rubies and those that are rubbish will not be a simple matter, particularly since the fitness effects of most alleles are very likely contextually dependent (i.e., neither unconditionally beneficial nor unconditionally deleterious).

A second issue of concern with respect to the practical meaning of measures of genetic variation for quantitative traits involves the degree to which gametic-phase disequilibria exist among QTL. Consider the extreme situation in which the effects of all genes can be characterized as + and –, and imagine the case in which all chromosomes have four genes in either the series + – + – or – + – +. This is a case of complete repulsion disequilibrium, in that every – gene is paired with a + gene. In such a population, every diploid individual would have exactly four + and four – genes, and there would be no expressed genetic variation for the trait, despite the presence of substantial variation within each locus. The expected existence of such repulsion disequilibrium was first suggested by Mather (1942) in his polygenic balance model for characters under stabilizing selection and has been formally verified in a number of theoretical studies (e.g., Bulmer, 1971; Gavrilets and Hastings, 1995). The alternative extreme is the situation in which all chromosomes are of state + + + + or – – – –. In this case, disequilibrium is of the coupling (or reinforcing) type, and the level of expressed genetic variance for the trait will be exaggerated with respect to the equilibrium situation in which + and – genes are distributed randomly. The point of this example is that measurable (expressed) genetic variance for quantitative traits can deviate substantially from the potential level that would be observed in the absence of disequilibria. If significant repulsion disequilibrium exists for quantitative traits under stabilizing selection, then a pool of hidden genetic variance will exist that can subsequently be released through recombination. Unless linkage is extraordinarily strong, such a reserve of genetic variation may be made available to a population exposed to directional selection much more rapidly than new variation can be created by mutation.

Only a few attempts have been made to evaluate whether substantial disequilibria exist for quantitative traits. Studies with *Daphnia* have yielded evidence of both coupling and repulsion disequilibria for life history traits (Lynch, 1984; Lynch and Deng, 1994). A series of comparisons of intact chromosomes extracted from wild populations of *Drosophila* with recombinants between them also have suggested the existence of coadapted groups of linked alleles (reviewed in Lynch and Walsh, 1998). Future studies that simply removed populations from the field and subjected them to several rounds of recombination with a minimum of selection could provide substantial insight into this issue. If significant repulsion disequilibria exist, then the level of expressed genetic variance for a quantitative trait should increase following a few rounds of random mating.

All these examples serve to illustrate that although we are reasonably good at quantifying standing levels of genetic variance for quantitative characters, we still are far from the point of being able to make unequivocal statements about the relevance of such variation for evolution, other than the fact that (on average) characters with higher levels of additive genetic variance will probably respond more strongly to short-term evolutionary pressures. We still are unable to make any general statements as to how rapidly organisms can evolve in response to long-term selective pressures. This question is of substantial importance to breeders of economically important species (Hill, 1982). It also is of central importance in conservation biology, especially with respect to issues such as global warming. In principle, populations can cope with environmental change by simply evolving *in situ*, but there is a cost to natural selection in the form of reduced survival and/or fecundity. Thus, there must be a critical rate of environmental change beyond which the cost of natural selection is so great that populations cannot maintain themselves demographically. By definition, this point is equivalent to the maximum sustainable rate of evolution, and any long-term ecological changes that exceed this rate must spell ecological catastrophe for the species in question. Understanding the limits to the rate of evolution is a substantial challenge for applied evolutionary biology. Recent theory has provided some qualitative insight into the issue (Lynch and Lande, 1993; Burger and Lynch, 1994), but convincing quantitative answers will require empirical work.

## POPULATION GENETICS AND POPULATION DYNAMICS

One of the most glaring gaps in the broad field of population biology is the lack of integration between its two subdisciplines: population genetics and population ecology. Historically, population geneticists have assumed populations of either infinite size or of constant finite size, while population ecologists usually have ignored genetics entirely. As a consequence, we know very little, from either the theoretical or the empirical perspective, of the relationship between population genetics and population dynamics/population size. In principle, one might expect that populations that evolve adaptively also would expand numerically, and that populations that find themselves in situations where deleterious genes have accumulated would decline numerically. However, the exact nature of such relationships presumably also depends on the extent to which interacting species (competitors, predators, parasites, and prey) are undergoing evolutionary adaptation versus deterioration. Moreover, important synergisms between population size and evolutionary dynamics may be mediated by random genetic drift, in that small populations are more vulnerable to accumulation of deleterious mutations and less likely to fix beneficial mutations than are large populations.

Theoretical work in this area has been largely confined to a single issue of interest to conservation biologists: the critical effective size below which a population is evolutionarily compromised (reviewed in Lande, 1995; Lynch, 1996; Burger and Lynch, 1997). It generally is believed that a population will progressively accumulate a deleterious mutation load by random genetic drift if its effective size remains much smaller than the reciprocal of the average deleterious effect of a new mutation. Eventually, this load may become so great that the population cannot maintain itself, at which point a mutational meltdown rapidly leads to population size reduction, an enhanced rate of accumulation of deleterious mutations by random drift, and finally to population extinction. However, there must be an approximate upper bound to the effective size of a population beyond which the population behaves adaptationally as though it were effectively infinite.

Quantitative values for these critical effective population sizes are highly dependent on the distribution of mutational effects for quantitative characters, an issue that is currently the subject of substantial debate (Keightley, 1994; Crow, 1997; Garcia-Dorado, 1997). The best we can say at this point is that populations with long-term effective sizes smaller than 100 individuals are likely to be doomed evolutionarily, whereas populations with long-term effective sizes in excess of 1,000 or so individuals are unlikely to be greatly compromised evolutionarily.

## ISSUES CONCERNING OUTBREEDING DEPRESSION AND SPECIES INCOMPATIBILITIES

A fundamental issue in evolutionary genetics with broad implications is the extent to which genomes from distantly related populations are mutually compatible. For example, in conservation biology, there is considerable concern about the consequences of hybridization between introduced and native populations (e.g., the stocking of salmonid hatcheries with members of foreign drainage basins). In addition, attempts to understand the mechanisms of speciation can hardly proceed without giving attention to the issue of genomic incompatibility.

What have we learned over the past few decades about outbreeding depression? First, a number of well-established examples demonstrate that crosses between closely related populations can give rise to depressed fitness in the $F_1$ progeny (Templeton *et al.*, 1976; Burton, 1990; Waser and Price, 1994). Second, the few cases that have attempted to laboriously dissect the genetic basis of species differentiation (all with *Drosophila*) have shown that a very large number of factors are involved, even when very closely related taxa are considered (Wu and Palopoli, 1994). In both types of studies, it is clear that negative epistatic interactions between foreign gene pools are involved.

What has not emerged is any general understanding of the genes (or even gene families) that are involved in hybrid breakdown, and it is perhaps unreasonable to expect that any general statements ever will be possible in this area. Isolated populations will diverge over time as they randomly go to fixation for mutations arising at different loci. Some aspects of divergence are certainly due to adaptation to local environmental differences, while others very likely arise as a consequence of intrinsic adaptation (as new mutant alleles are selected for or against on the basis of the genetic background in which they arise). With polygenic systems, there can be innumerable solutions to the same evolutionary problem. Thus, because of the stochastic nature of the mutation and fixation processes, it seems very likely that the genes that underlie the reproductive barriers between any particular pair of species will be largely unique to that species pair and of little general relevance.

A more general concern is the relevance of genes that can be shown to be involved in reproductive incompatibilities to the process of speciation even in particular pairs of species. The issue here is that a large fraction of the genetic differences between even the most closely related species generally will have arisen subsequent to the time at which reproductive incompatibility was first established. Like the issue of genetic variance, the problem here is that the genes of greatest relevance to the process of speciation may be "rubies in the rubbish."

In principle, this type of problem can be minimized by restricting studies of speciation to extremely closely related taxa (or perhaps more suitably, to distantly related populations of the same species), but even this does not eliminate the issue of the specificity of genetic differentiation. This is not to say that no general statements about the mechanisms of reproductive isolation ever will be possible. One such generalization has been long established for species with sex chromosomes: the observation, known as Haldane's rule, that when only one sex resulting from a hybridization event exhibits reduced survivorship and/or fertility, it is invariably the heterogametic sex (Laurie, 1997; Orr, 1997).

## SUMMARY

The year 1998 marked the 80th anniversary of Fisher's (1918) insightful paper, a compelling mathematical unification of various observations on quantitative characters, statistics, and basic Mendelian mechanisms of inheritance. Despite the major mathematical advances that have been made since then, quantitative genetic theory is still framed largely in statistical–phenomenological terms rather than in terms that describe the mode of gene action. Part of the reason for this state of affairs is our extremely shallow understanding of the genetic basis of quantitative trait variation and of the mechanisms by which complex characters evolve. Although a vast number of empirical studies of quantitative characters have been performed with laboratory populations, with species of agronomic importance, and with humans, most of these have been focused on population-level properties, such as means and variances. New molecular techniques promise to yield important insights into the genetic basis of variation for specific traits in a few model species in the very near future, but the extent to which molecular dissection will provide a complete resolution to our understanding of the evolutionary process remains to be seen. This will depend largely on the degree to which quantitative trait variation is defined by a large number of genes of very small effects, an issue for which there is a fair amount of circumstantial evidence in a number of cases. Although recent work suggests that a few genes of moderate effect often can explain a substantial fraction of the standing variation for quantitative traits, it remains to be seen whether this fraction is of much relevance to the process of adaptive evolution. If a large fraction of the variation that contributes to adaptive divergence is attributable to genes of very small effects, then molecular analysis of QTLs may be of limited value to our understanding of the evolution of natural populations (although it will certainly still play a major role in artificial selection programs).

The ability to evolve is just as much a defining attribute of life as the ability to replicate. Thus, it is remarkable that while ecologists have contributed greatly to our understanding of the chemical, physical, and biological limits that constrain the spatial and temporal distributions of organisms, evolutionary biologists have given almost no attention to the evolutionary limits of species (i.e., to the abilities of species to adapt evolutionarily to selective challenges as opposed to going extinct). If we agree that the global environment is changing at an unprecedented rate, then it should be clear why the development of a deep understanding of this issue should be one of the central challenges of evolutionary biology. For the past three decades, ecologists have pursued a large number of large-scale and long-term programs of ecosystem research. Because of the stochastic and very long-term nature of evolutionary processes, it is difficult to see how a mechanistic understanding of evolution can be achieved without an empirical agenda that embraces highly replicated, multigenerational studies.

## REFERENCES

Bulmer, M. G., 1971, The effects of selection on genetic variability, *Am. Natur.* **105:**210–211.

Burger, R., and Lynch, M., 1994, Evolution and extinction in a changing environment: A quantitative genetic analysis, *Evolution* **49:**151–163.

Burger, R., and Lynch, M., 1997, Adaptation and extinction in changing environments, in: *Environmental Stress, Adaptation, and Evolution* (R. Bijlsma and V. Loeschcke, eds.), pp. 209–240, Birkhauser Verlag, Basel.

Burton, R. S., 1990, Hybrid breakdown in developmental time in the copepod *Tigriopus californicus*, *Evolution* **44:**1814–1822.

Charlesworth, B., Lande, R., and Slatkin, M., 1982, A neo-Darwinian commentary on macroevolution, *Evolution* **36:**474–498.

Cheverud, J. M., and Routman, E. J., 1995, Epistasis and its contribution to genetic variance components, *Genetics* **139:**1455–1461.

Crow, J. F., 1992, Mutation, mean fitness, and genetic load, *Oxford Surv. Evol. Biol.* **9:**3–42.

Crow, J. F., 1997, The high spontaneous mutation rate: is it a health risk? *Proc. Natl. Acad. Sci. USA* **94:**8380–8386.

Doebley, J., and Stec, A., 1993, Inheritance of the morphological differences between maize and teosinte: Comparison of results for two $F_2$ populations, *Genetics* **134:**559–570.

Eizinger, A., and Sommer, R. J., 1997, The homeotic gene *lin-39* and the evolution of nematode epidermal cell fates, *Science* **278:**452–454.

Fisher, R. A., 1918, The correlation between relatives on the supposition of Mendelian inheritance, *Trans. R. Soc. Edinburgh* **52:**399–433.

Garcia-Dorado, A., 1997, The rate and effects distribution of viability mutation in *Drosophila*: Minimum distance estimation, *Evolution* **51:**1130–1139.

Gavrilets, S., and Hastings, A., 1995, Dynamics of polygenic variability under stabilizing selection, recombination, and drift, *Genet. Res.* **65:**63–74.

Gerhart, J., and Kirschner, M., 1997, *Cells, Embryos, and Evolution*, Blackwell Science, Inc., Malden, Massachusetts.

Hazel, L. N., 1943, The genetic basis for constructing selection indices, *Genetics* **28:**476–490.

Hill, W. G., 1982, Rates of change in quantitative traits from fixation of new mutations, *Proc. Natl. Acad. Sci. USA* **79:**142–145.

Hollocher, H., Templeton, A. R., DeSalle, R., and Johnston, J. S., 1992, The molecular through ecological genetics of abnormal abdomen in *Drosophila mercatorum*. III. Components of genetic variation in a natural population of *Drosophila mercatorum*, *Genetics* **130:**355–366.

Houle, D., Morikawa, R., and Lynch, M., 1996, Comparing mutational variabilities, *Genetics* **143:**1467–1483.

Keightley, P. D., 1994, The distribution of mutation effects on viability in *Drosophila melanogaster*, *Genetics* **138:**1315–1322.

Lai, C., Lyman, R. F., Long, A. D., Langley, C. H., and Mackay, T. F. C., 1994, Naturally occurring variation in bristle number associated with DNA sequence polymorphisms at the *scabrous* locus of *Drosophila melanogaster*, *Science* **266:**1697–1702.

Lande, R., 1988, Quantitative genetics and evolutionary theory, in: *Proceedings of the Second International Conference on Quantitative Genetics* (B. S. Weir, E. J. Eisen, M. M. Goodman, and G. Namkoong, eds.), pp. 71–84, Sinauer Associates, Sunderland, Massachusetts.

Lande, R., 1995, Mutation and conservation, *Cons. Biol.* **9:**782–791.

Laurie, C. C., 1997, The weaker sex is heterogametic: 75 years of Haldane's rule, *Genetics* **147:**937–951.

Lowe, C. J., and Wray, G. A., 1997, Radical alterations in the roles of homeobox genes during echinoderm evolution, *Nature* **389:**881–883.

Lynch, M., 1984, The limits to life history evolution in *Daphnia*, *Evolution* **38:**465–482.

Lynch, M., 1996, A quantitative genetic perspective on conservation issues, in: *Conservation Genetics: Case Histories from Nature* (J. Avise and J. Hamrick, eds.), pp. 471–501, Chapman and Hall, New York.

Lynch, M., and Deng, H.-W., 1994, Genetic slippage in response to sex, *Am. Natur.* **144:**242–261.

Lynch, M., and Lande, R., 1993, Evolution and extinction in response to environmental change, in: *Biotic Interactions and Global Change* (P. Kareiva, J. Kingsolver, and R. Huey, eds.), pp. 234–250, Sinauer Associates, Sunderland, Massachusetts.

Lynch, M., and Walsh, B., 1998, *Genetics and Analysis of Quantitative Traits*, Sinauer Associates, Sunderland, Massachusetts.

Mackay, T. F. C., and Langley, C. H., 1990, Molecular and phenotypic variation in the *achaete-scute* region of *Drosophila melanogaster*, *Nature* **348:**64–66.

Mather, K., 1942, The balance of polygenic disorders, *J. Genetics* **43:**309–336.

Orr, H. A., 1997, Haldane's rule, *Annu. Rev. Ecol. Syst.* **28:**195–218.

Templeton, A. R., Sing, C. F., and Brokaw, B., 1976, The unit of selection in *Drosophila mercatorum*. I. The interaction of selection and meiosis in parthenogenetic strains, *Genetics* **82:**349–376.

Templeton, A. R., Hollocher, H., Lawler, S., and Johnston, J. S., 1989, Natural selection and ribosomal DNA in *Drosophila*, *Genetics* **31:**296–303.

Waser, N. M., and Price, M. V., 1994, Crossing-distance effects in *Delphinium nelsonii*: Outbreeding and inbreeding depression in progeny fitness, *Evolution* **48:**842–852.

Wu, C.-I., and Palopoli, M. F., 1994, Genetics of postmating reproductive isolation in animals, *Annu. Rev. Ecol. Syst.* **27:**283–308.

13

# Genetics of Species Differentiation

## What Is Unknown and What Will Be Unknowable?

CHUNG-I WU

## MUSINGS ABOUT GENERALITY IN BIOLOGY

What is the genetic nature underlying species differences? What are the forces driving the genetic divergence? As evolutionists, we may be as much interested in the diversity of life as in, say, the universality of certain biological properties. It will be stated at the outset what I consider may be the limit of our ability to understand the nature of diversity and the forces shaping it. In science, we strive to deduce the generality of the scheme of things. Physical sciences have been phenomenally successful in this endeavor. Newtonian mechanics applies equally well to planetary motion, the falling of apples, and the production of sound. The more general a principle is, the more intellectually profound it ought to be. Is it possible that in biology and especially in evolutionary biology the more general a principle is, the less profound it would seem? The "limit" I refer to is the paucity of general principles that are intellectually challenging.

When is generality less satisfying than the specific details? Let us compare the diversity of life to a vast collection of chess games with each lineage of life being equated to a game. Of course, the most general

CHUNG-I WU • Department of Ecology and Evolution, University of Chicago, Chicago, Illinois 60637.

*Evolutionary Biology, Volume 32*, edited by Michael T. Clegg *et al.*
Kluwer Academic / Plenum Publishers, New York, 2000.

principles of chess are the patterns of movement, the capture of opposing pieces, and the end of a game. An intelligent person who has never been exposed to chess would probably figure out all the general principles of the game after many rounds of observation. Would understanding the rules of the game be viewed as being at a higher intellectual plane than analyzing the peculiarity and idiosyncrasy of a classic match between the masters? What if the true intellectual value of evolutionary biology lies in the peculiarity? Like a collection of classic chess matches, specific biological systems may share some features, but these features are often only partially generalizable. Exceptions are expected to be the rule and generalities may be trivial, compared with the idiosyncrasies. Among recent publications, the worst offense that puts generality at the pinnacle of intellectual pursuit is John Horgan's book *The End of Science* (1996). Horgan entitled each chapter, The End of (Physics, Cosmology, Evolutionary Biology, and so on). In his reasoning, once we discovered the general principle of evolution by natural selection, the best of evolutionary biology would be already behind us. It had indeed been said many times before Horgan that we are merely filling in the "pigeon holes" in the large intellectual framework erected by Charles Darwin (e.g., Lewontin, 1974). Following that logic, we might suggest that the most general principle for making money in a stock market ("to buy low and sell high") is a profound idea. Instead, the real challenge and excitement is in the details of individual products, their market milieu, and numerous other specific elements.

What is stated above may apply particularly well to the study of speciation and species differentiation where the essence is in the specific. After all, if two populations are separated long enough, they will eventually become distinct species (and even this simple rule may not be quite that general if sympatric speciation occurs occasionally). It is the specific details of the process of speciation between any given pair of diverging species, a process of differential adaptation, that intrigue us. Recently, speciation studies have reached the molecular level, and the nature of genes involved is as diverse as the biology of the species studied [e.g., abalone (Swanson and Vacquier, 1998) vs. *Drosophila* (Ting *et al.*, 1998)]. We may expect even greater variety and less commonality from the molecular studies of speciation in the future.

Below, I shall use two cases of interspecific studies to elaborate on the theme. In case 1, the attempt to generalize (or overgeneralize) had in fact impeded our understanding of an important phenomenon in species hybridization until recently. Case 2 is an example of the uncertainty in generalization across different scales. Unlike in physical systems where principles that apply to a smaller scale usually work in a larger one (e.g., in astrophysics or geology), in evolutionary genetics, extrapolation from within-species studies to between-species observations may be off the mark by a wide margin. In that sense, it is akin to trying to understand the long-term trend of a stock market by extrapolating from the analysis of the one-day movement. The limit to what we may know is a consequence of the dimensionality of the system and the impossibility of being in all the dimensions at the same time to conduct the analysis.

## CASE 1: HALDANE'S RULE AND THE PERIL OF OVERGENERALIZATION

So, what possibly can be wrong with our attempt to generalize? One may argue that the worst that could happen is that we fail to find the generality we hope for. However, if a general pattern does not exist, the search for it may mislead us into forgoing a true understanding in a more restricted setting. A recent example is our attempt to understand Haldane's rule. This rule refers to the observations that, in interspecific crosses, the heterogametic hybrids tend to suffer inviability or sterility more so than the homogametic sex (Haldane, 1922). An obvious explanation attributed to Muller (1940) is that $F_1$ hybrids of the heterogametic sex have the X chromosome from one species only and the autosomes from both species, whereas the homogametic sex has the complete haploid genomes, one set from each species. In *Drosophila* and mammals, hybrid males thus are imbalanced. This explanation, having been accepted for 50 years, was tossed out in the mid-80s because specific cases for which Muller's explanation does not hold true are common (e.g., hybrid male sterility in *Drosophila*). Given those results, we had two options: (1) To reject Muller's idea and renew our attempt to find a truly general explanation; this became the accepted practice for many years, and several different explanations were proposed, but none was better than the original one; or (2) alternatively, we might question whether Haldane's rule is a unitary phenomenon and whether a general explanation exists. If one takes this route, then an additional explanation for male sterility in *Drosophila* and mammals must be found. In fact, a cursory look at *Drosophila* and mammalian data lends itself to the interpretation that hybrid male sterility must be evolving at an extraordinarily high rate. This rapid evolution is independent of Muller's mechanism (Muller and Pontecorvo, 1942). The reason that rapid evolution of hybrid male sterility never has been suggested to be a component of

Haldane's rule is that it does not account for the data from bird and butterfly studies.

The subject has been extensively, maybe too extensively, reviewed (Wu and Davis, 1993; Wu *et al.*, 1996; Orr, 1997; Laurie, 1997). Now a consensus appears to have been reached: Muller's mechanism and the rapid evolution of hybrid male sterility are two separate causes for hybrid incompatibility, which obeys Haldane's rule. [There are in fact several other forces operating (Wu *et al.*, 1996).] Neither explanation is sufficient for all cases, and in many cases the two explanations are antagonistic. The quest for a general explanation had resulted in the rejection of both. In fact, one may argue that the rapid evolution toward hybrid male sterility is a much more interesting observation than Haldane's rule itself. The data existed but were not properly interpreted (because the observation does not seem to be completely general). The implications of this phenomenon in relation to sexual selection driving the evolution of male reproductive characters (Eberhard, 1985) have been discussed elsewhere (Wu *et al.*, 1996; Hollocher *et al.*, 1997; Wu and Hollocher, 1998). The lesson is that before embarking on the search for a general solution to a problem we need to have *a priori* reasons to think that it exists. Like the proverbial black cat in a dark room, catching it may be difficult, but the task would be much more difficult if we do not even know whether it is there.

The debate on Haldane's rule is instructive in another respect. In the quest for generality, the more knowledge we have about a particular system, the easier it is to consider the system a special case and any property of the system would be deemed less amenable for extrapolation. Since the bulk of the data came from *Drosophila* studies and so much is known about them, many explanations based on *Drosophila* data were considered "too specific." For example, the peculiar genetic control of spermiogenesis with the near-complete absence of transcriptional regulation (Wu and Davis, 1993) rarely has been considered, in part due to the lack of understanding of spermatogenesis in other species groups. This dilemma of "knowing too much" may apply to other topics. Imagine if a locus with a tremendous amount of nucleotide polymorphisms were revealed to population geneticists in the 1970s and every aspect of the observations conformed to the "balanced school" of thought. Since we did not choose this locus based on any known biology, we might feel justified in assuming that it was a random locus. The results might be cited (perhaps widely) to favor the balanced school over the classic–neutral camp. However, if it were later revealed that the locus is a major histocompatibility complex (MHC) gene, we might downgrade the results to a special case. The obvious irony is that we could be more confident about the generality of data, if we know less about the underlying biology.

# CASE 2: SPECIES DIFFERENTIATION AND THE UNCERTAINTY IN EXTRAPOLATION FROM POPULATION GENETIC VARIATION

> The essence of speciation is the production of two well-integrated gene complexes from a single parental one. (Mayr, 1963)
>
> My own speculations on evolution were dominated by the thought . . . of coadaptive interaction systems as wholes. (Wright, 1982)
>
> Such terms as integrated coadapted complex, . . . cohesion of the gene pool, . . . were more rhetorical than scientific. (Kimura, 1983)

In every branch of science, the attempt to understand a larger system usually is reduced to a simpler task of understanding a smaller, more manageable model system first. What is learned in the smaller model is then extrapolated to the larger system. This certainly is true in the attempt to learn about species differentiation. Instead of analyzing species differentiation directly, evolutionists have resorted to analyzing variation within species because that is where genetics can be done and population genetics theories developed. In the first chapter of *The Origin of Species*, "Variation under Domestication," Darwin (1859) gave the obvious reason for studying population genetic variations, because they are the ultimate source of evolution, giving rise to the biological diversity we strive to understand. We may say that population genetic variation is interesting ultimately in the context of speciation and species differentiation. A second and equally important reason for studying population genetic variations is to understand the forces that pattern them.

Below, I shall first discuss the recent understanding of the genetics of species differentiation and then address the issues of extrapolation from population genetic analysis to account for species differentiation. Before the neo-Darwinian synthesis, it had been common to speculate that the divergence at the species level is of a very different category than variations within species. While the rationale and empirical basis for this speculation have been strongly rebuked by the new synthesis, it nevertheless is far from safe to interpret the genetics of species divergence by extrapolating from population genetic variations.

To begin with, what do we know about the genetics of species differentiation? Can we even guess how many genes it usually takes to make a new species? Let us consider the difference between human and chimpanzee, currently estimated to be about 1.7% at the nucleotide level (Sibley and Ahlquist, 1984; Koop *et al.*, 1989; Li, 1997). However, what does the observation that human and chimpanzee have diverged by 1.7% really tell us about the genes involved in our own differentiation from the chimpanzee? The relatively small difference of 1.7% represents about 50 million base pairs in a genome of $3 \times 10^9$ base pairs. How many base pairs have to be changed in order to convert a chimpanzee embryo into a human in most of the morphological, physiological, and behavioral senses? Our ignorance about the genetics of species differences becomes quite obvious here. We cannot refute the view that 50 of the most critical base pair changes could convert a chimpanzee into a passable human being. Neither can we rule out the extreme opposite view that 500,000 base pairs, about 1% of the total differences, would be needed. Obviously, our ignorance about the genetics of species difference spans many orders of magnitude.

To tackle this problem, one can go directly to studying interspecific genetics. Unfortunately, every trait of interest has a complex genetic basis, and the tools and materials are severely limited. Only in the last 5 years did we have a rough sketch of the general architecture of the genetics of species differentiation. If *Drosophila* is used as a guide, the biological differences between human and chimpanzee would likely involve a large number of base pair changes that is orders of magnitude higher than 50 base pairs. Between even the most closely related species of *Drosophila*, the divergence in the genetic program of spermatogenesis (which is manifested as hybrid male sterility) amounts to at least 120 loci (Cabot *et al.*, 1994; Palopoli and Wu, 1994; Wu *et al.*, 1996; True *et al.*, 1996; Davis and Wu, 1996; Hollocher and Wu, 1996; Hollocher *et al.*, 1997; Wu and Hollocher, 1998). What is most relevant to this short note is the extent of fitness interaction among these loci. The fitness consequence of each allelic substitution can be close to 100% or 0%, depending on its interaction with some conspecific loci. We can talk about the fitness effect of a particular genetic change only in the context of a specified background, but we often project the fitness consequence to the neighboring species, as is apparent in many population genetic tests. During this long divergence, there may be thousands, if not millions, of base pair changes. However, we often make this stringent assumption that, for example, a Phe to Leu change at a certain position of a particular protein is associated with a fitness difference, $s$, and this value remains constant while the rest of the genome undergoes thousands or millions of changes.

I shall discuss the uncertainty in extrapolating from within-species observations to interspecific differences, using one particular test as an example. The caveat probably applies to other tests as well. Fitness epistasis, if prevalent, would make it difficult to interpret the molecular divergence between species. As a consequence of ever-changing genetic backgrounds, selective pressure would be capricious, and thus difficult to quantify. For a linear projection to work, the selective pressure has to be constant within the space and time under consideration. It was this assumption that Mayr (1959) questioned, rather than the usefulness of mathematics in population genetics, as Haldane (1964) defended.

Whether selection can be assumed to be constant is relevant to the comparison of within-species DNA polymorphism with between-species DNA divergence (MacDonald and Kreitman, 1991). Briefly, mutations under positive directional selection are fixed rapidly in the population; such mutations have very short sojourn times in the population relative to neutral fixations. If amino acid replacements are often under the influence of positive selection, there should be many more such replacements between species than would be predicted based on the level of amino acid polymorphism observed within species. The ratio of amino acid replacements between species to those within species by itself is not informative unless we also know the relative time scale of between- versus within-species divergence. MacDonald and Kreitman (1991) used silent changes to provide the relative time scale and devised an appropriate statistical test (see Sawyer and Hartl, 1992). They noted a significant excess of amino acid replacements between the *Adh* genes of different *Drosophila* species, compared with the expected value extrapolated from the level of amino acid polymorphism within species.

The most logical and parsimonious explanation for the significant excess is positive selection favoring many amino acid replacements since the species diverged. This pattern subsequently has been observed in several genes surveyed. Inherent in the MacDonald and Kreitman's (1991) test, as well as many population genetic models, is the linear extrapolation from within-species polymorphisms to between-species divergence. In other words, selective pressure remains constant in the period being studied. Without the assumption of constant selection, the excess in amino acid substitution between species can be accounted for by fluctuating negative selection (less negative selection in the past). This is what Takahata (1988) refers to as "fluctuating neutral space."

Constant positive selection, however, is a double-edge sword; if it can lead to an amino acid replacement in one species, it may do so also in the other. Consequently, there would be no observable adaptive divergence. When an amino acid replacement has a *constant* positive fitness in the two species lineages since their divergence, the fitness value has to be quite small for adaptive divergence to occur. An adaptive mutation with a fitness advantage of $s = 1\%$ would have a 2% chance of becoming fixed in the population (Crow and Kimura, 1970, p. 426); hence, each species only needs to generate 228 mutations, from the time of speciation until the recent past, to have a 99% ($1–0.98^{228}$) chance of becoming fixed with this adaptive mutation in the population. For species like *Drosophila*, which have a relative large population size and short generation time, mutations with a constant 1% advantage should all have been fixed in both species a long time ago. The value of $s$ therefore cannot be larger than a threshold, determined by the population size, generation length, and the divergence time of the species, above which convergent evolution is near certainty. On the other hand, $s$ has to be larger than 1/2 N to overpower genetic drift. I calculate that the range of $s$ has to fall within one order of magnitude for differential fixation of positively selected variants to be a valid explanation of *Drosophila* data. Therefore, while the excess of amino acid replacements between species is clearly incompatible with a constant neutral space [i.e., neutrality under constant negative selection (Takahata, 1988)], the constant positive selection model also may seem unduly restrictive.

It then may be appropriate to abandon the parsimonious assumption of constant selection. The dilemma is that without such an assumption the data can be compatible with several interpretations. Neutralism always has allowed some degree of variable negative selection. An excess of amino acid replacements between species may be explained by negative selection that was less stringent sometime in the past, allowing more amino acid changes. The mechanism of relaxation of negative selection has been invoked to explain elevated amino acid substitution rates after gene duplication (Li, 1997; Zhang *et al.*, 1998). Takahata (1988) also suggested fluctuating neutral space as a basis for the overdispersed molecular clock. To account for the observed excess of amino acid replacements between species by fluctuating negative selection, a significant portion of the whole protein must be involved, whereas fewer sites need to experience changing selective pressure under the positive selection model. Whether the positive selection model should be accepted on this argument alone is debatable. In certain cases, when one biochemical pathway became less used than the alternate, selective constraints on the whole proteins of that pathway may have become relaxed (e.g., Eanes *et al.*, 1993).

Is it the environment or the genetic background that contributes to the fluctuation in selective pressure? These are not mutually exclusive factors. Certainly, the genetic background would have profound effects on the fitness consequences of individual genetic changes. An explicit model is the concept of covarions (Fitch and Markowitz, 1970), which suggests the existence of a subset of amino acid sites within a protein that covary evolutionarily. The composition of covarying sites changes from time to time as the consequence of previous amino acid replacements. Therefore, variable sites of the same gene in different taxa may be distributed differently along the gene.

## CONCLUSIONS

The intellectually challenging aspects of evolutionary biology often may be idiosyncratic (or at least not highly general), whereas the generality may be rather trivial. In the first case discussed, it is shown that the attempt to force a general principle on Haldane's rule had in fact impeded a true understanding of the complex phenomena. The second case is about the attempt to extrapolate from short-term evolution (i.e., population genetic variations) to the longer-term changes (the divergence between species). This may be another example of reductionism failing to extrapolate from the study of a simpler system to the understanding of a more complex one.

## REFERENCES

Cabot, E. L., Davis, A. W., Johnson, N. A., and Wu, C.-I., 1994, Genetics of reproductive isolation in the in the *Drosophila simulans* clade: Complex epistasis underlying hybrid male sterility, *Genetics* **137:**175–189.

Crow, J. F., and Kimura, M., 1970, *An Introduction to Population Genetics Theory*, Harper and Row, New York.

Darwin, C., 1859, *The Origin of Species*, John Murray, London.

Davis, A. W., and Wu, C.-I., 1996, The broom of the sorcerer's apprentice: The fine structure of a chromosomal region causing reproductive isolation between two sibling species of *Drosophila*, *Genetics* **143:**1287–1298.

Eanes, W. F., Kirchner, M., and Yoon, J., 1993, Evidence for adaptive evolution of the *G6pd* gene in the *Drosophila melanogaster* and *D. simulans* lineages, *Proc. Natl. Acad. Sci. USA* **90:**7475–7479.

Eberhard, W. G., 1985, *Sexual Selection and Animal Genitalia*, Harvard University Press, Cambridge, Massachusetts.

Fitch, W. M., and Markowitz, E., 1970, An improved method for determining codon variability in a gene and its application to the rate of fixation of mutations in evolution, *Biochem. Genet.* **4:**579–593.

Haldane, J. B. S., 1922, Sex ratio and unisexual sterility in hybrid animals, *J. Genet.* **12:**101–109.

Haldane, J. B. S., 1964, A defense of beanbag genetics, *Persp. Biol. Med.* **7:**343–359.

Hollocher, H., and Wu., C.-I., 1996, The genetics of reproductive isolation in the *Drosophila simulans* clade: X vs. autosomal effects and male vs. female effects, *Genetics* **143:**1243–1255.

Hollocher, H., Ting, C. T., Pollack, F., and Wu, C.-I., 1997, Incipient speciation by sexual isolation in *Drosophila melanogaster*: Variation in mating preference and correlation between sexes, *Evolution* **51:**1175–1181.

Horgan, J., 1996, *The End of Science*, Broadway Books, New York.

Kimura, M., 1983, *The Neutral Theory of Molecular Evolution*, Cambridge University Press, Cambridge, England.

Koop, B. F., Tagle, D. A., Goodman, M., and Slightom, J. L., 1989, A molecular view of primate phylogeny and important systematic and evolutionary questions, *Mol. Biol. Evol.* **6:**580–612.

Laurie, C. C., 1997, The weaker sex is heterogametic: 75 years of Haldane's rule, *Genetics* **147:**937–951.

Lewontin, R. C., 1974, *The Genetics Basis of Evolutionary Changes*, Columbia University Press, New York.

Li, W. H., 1997, *Molecular Evolution*, Sinauer Associates, Sunderland, Massachusetts.

Mayr, E., 1959, Where are we? *CSH Sympos. Quant. Biol.* **24:**1–14.

Mayr, E., 1963, *Animal Species and Evolution*, The Belknap Press, Cambridge, Massachusetts.

McDonald, J. H., and Kreitman, M., 1991, Adaptive protein evolution at the *Adh* locus in *Drosophila*, *Nature* **351:**652–654.

Muller, H. J., 1940, Bearing of the *Drosophila* work on systematics, in: *The New Systematics* (J. S. Huxley, ed.), pp. 185–268, Clarendon Press, Oxford, England.

Muller, H. J., and Pontecorvo, G., 1942, Recessive genes causing interspecific sterility and other disharmonies between *Drosophila melanogaster* and *simulans*, *Genetics* **27:**157.

Orr, H. A., 1997, Haldane's rule, *Annu. Rev. Ecol. Syst.* **28:**195–218.

Palopoli, M., and Wu, C.-I., 1994, Genetics of hybrid male sterility between *Drosophila* sibling species: A complex web of epistasis is revealed in interspecific studies, *Genetics* **138:**329–341.

Sawyer, S. A., and Hartl, D. L., 1992, Population genetics of polymorphism and divergence, *Genetics* **132:**1161–1176.

Sibley, C. G., and Ahlquist, J. E., 1984, The phylogeny of the hominoid primates, as indicated by DNA–DNA hybridization, *J. Mol. Evol.* **20:**2–15.

Swanson, W. J., and Vacquier, V. D., 1998, Concerted evolution in an egg receptor for a rapidly evolving abalone sperm protein, *Science* **281:**710–712.

Takahata, N., 1988, More on the episodic clock, *Genetics* **118:**387–388.

Ting, C. T., Tsaur, S. C., Wu, M. L., and Wu, C.-I., 1998, A rapidly evolving homeobox at the site of a hybrid sterility gene, *Science* **282:**1501–1504.

True, J. R., Weir, B. S., and Laurie, C. C., 1996, A genome wide survey of hybrid incompatibility factors by the introgression of marked segments of *Drosophila mauritiana* chromosomes into *Drosophila simulans*, *Genetics* **142:**819–837.

Wright, S., 1982, Character change, speciation and the higher taxa, *Evolution* **36:**427–443.

Wu, C.-I., and Davis, A. W., 1993, Evolution of postmating reproductive isolation: The composite nature of Haldane's rule and its genetic bases, *Am. Natur.* **142:**187–212.

Wu, C.-I., and Hollocher, H., 1998, Subtle is nature: The genetics of species differentiation and speciation, in: *Endless Forms: Species and Speciation* (D. Howard and S. Berlocher, ed.), Oxford University Press, Oxford, England.

Wu, C.-I., Johnson, N. A., and Palopoli, M. F., 1996, Haldane's rule and its legacy: Why are there so many sterile males? *Trends Ecol. Evol.* **11:**281–284.

Zhang, J., Rosenberg, H., and Nei, M., 1998, Positive Darwinian selection after gene duplication in primate ribonuclease genes, *Proc. Natl. Acad. Sci. USA* **95:**3708–3713.

# Conclusions

MICHAEL T. CLEGG

What can we conclude from this set of chapter about limits to knowledge in evolutionary genetics? The very question requires that we look into the future and imagine those areas that will be forever refrac-tory to future innovation. This is a daunting task, because we are participants in an era of unparalleled technical innovation in biology. We have seen biology answer questions that seemed beyond the bounds of knowledge a decade or two decades ago. We are confident that the future will replicate the recent past and produce answers to today's seemingly intractable problems. Many of the chapters in this volume reflect this view and reflect the problem of separating the merely difficult from the set of biologically interesting questions that lie beyond the bounds of knowledge.

There are two big questions in biology: How does the organism work and how did it get to be the way it is? The first question is the province of cell, developmental, and physiological biology (and all their specialized descendants), but it is also fundamental to understanding biological adaptation. The second question is the major concern of evolutionary biology. (I include the problem of prediction within this second category.) I think it is fair to say that the larger community of biologists is confident that we will understand how the organism works in exquisite detail. To illustrate how our grasp of this question has changed, I recall as an undergraduate student more than 35 years ago taking introductory biology and finding that the text included a section entitled "what is life." This section included a serious discussion of *elan vital* and the notion that understanding life might involve extraphysical explanations. It is difficult to imagine such a discussion today. We are absolutely committed to the notion that physically based and mechanistic explanations of how the organism works will be obtained. Certainly the recent progress of biology provides a strong basis for such confidence.

The second question of how did life get to its present state is more

---

MICHAEL T. CLEGG • Department of Botany and Plant Sciences, University of California, Riverside, California 92521 problematic (as is the corollary of how do we use our understanding in the prediction and management of biological systems). Our understanding of the past is rooted in the theory of evolution by natural selection. This is a statistical theory, but as noted in different ways by several authors, we have just one realization of the stochastic process of earthbound biological evolution to study, and it is not very informative to apply statistical means to the understanding of a single realization. It seems clear that our understanding of the past and our ability to predict the future will remain hazy and weakly defined. That is not to say that we cannot understand the broad outlines of biological history, but rather that we are unlikely to know the precise historical scenario.

Because of the statistical nature of genetic transmission and natural selection, our theories of the past and future are broad generalizations that do not provide a detailed explanation for the particular outcomes that we study as biologists. A place where this theme of the particular is especially apparent concerns the issue of biological interaction. Biological organisms clearly exhibit strong interactions at the molecular, developmental, and metabolic levels, so there is good reason to believe that interaction also is manifest at the statistical level, but our theoretical tools for studying biological systems are based on the minimization of interaction in the interests of generality and tractability. Several authors discuss the problem of biological interaction in different contexts. Because the biological system is complex, with thousands of actually or potentially interacting components across both the genetic and environmental dimensions, the ability to understand and predict the statistical behavior of these systems is limited.

Let me conclude with one final thought. The triumph of the physical sciences was to deduce a set of laws that govern the behavior of physical systems. This law-bound framework defines the possible. Rosenberg (Chapter 3) argued that evolution, and hence, all of biology, is governed by the law of natural selection and that this is the only law in biology. Despite the fact that the theory of evolution by natural selection is the major unifying law of biology, over the past 25 years, evolutionary genetics has focused a great deal of attention on nonselective causes for molecular evolution. This focus on neutrality has been of value because it has presented a clear and well-defined null hypothesis for statistical analysis, but it also has directed attention away from the central problem of biological adaptation. What a curious detour to have so much attention to be devoted to nonadaptive explanations for molecular diversity. Perhaps this detour reflected the social environment of the times. The notion of subjectivity in science appeared in several chapters in terms of Baysean priors or in terms of our collective judgment on what constitutes an interesting question. A limit to knowledge may simply be our belief systems and their role in determining the directions that our search for knowledge takes.

# Index